RAPTORS IN CAPTIVITY

Guidelines for Care and Management

DEDICATION

This book is dedicated to Marlys Bulander, a person who was devoted to improving the welfare of migratory birds through her long career with the USFWS, Region 3. Marlys had a kind, gentle heart and a commanding passion for protecting the rights of wild birds. Her genuine encouragement and support of this project was a great inspiration.

Marlys will be greatly missed by many people, but her compassion for people and birds will continue to guide us in making ethical choices about the wildlife with which we have the privilege to interact. Her spirit will forever soar with eagles in the heavens.

RAPTORS IN CAPTIVITY
Guidelines for Care and Management

Lori R. Arent
The Raptor Center, College of Veterinary Medicine at the University of Minnesota

hancock
house

ISBN-13: 978-0-88839-613-6
ISBN-10: 0-88839-613-9

Cataloging in Publication Data

Arent, Lori
   Raptors in Captivity: Guidelines for Care and Management/ Lori R. Arent.

Includes index.
ISBN 0-88839-613-9

   1. Captive wild birds. 2. Birds of prey. I. Title.

QL677.78.A73 2006   639.9'789   C2006-902551-7

Editing: Theresa Laviolette
Production: Mia Hancock
Image Editing: Laura Michaels
Cover design: Mia Hancock

Published simultaneously in Canada and the United States by

HANCOCK HOUSE PUBLISHERS LTD.
19313 Zero Avenue, Surrey, B.C. Canada V3S 9R9
(604) 538-1114 Fax (604) 538-2262

HANCOCK HOUSE PUBLISHERS
1431 Harrison Avenue, Blaine, WA U.S.A 98230-5005
(604) 538-1114 Fax (604) 538-2262

Website: **www.hancockhouse.com**
Email: **sales@hancockhouse.com**

Contents

List of Figures

Chapter 6   Maintenance Care

Chapter 7   Medical Care

Chapter 8   Training

Chapter 9   Transport

Chapter 10   Recovering a Lost Bird

Appendix C Raptor Enclosures

List of Tables

Preface

Enraptured with raptors. People around the world hold a fascination for hawks, owls, eagles, falcons, osprey, and vultures; an attraction that has no doubt preceded written history. To the ancient Greeks, the owl was the companion of Pallas Athena, the goddess of wisdom, and was a protector accompanying Greek armies to war. The ancient Romans and Egyptians, as well as many other civilizations, also incorporated owls, eagles, and falcons into their mythology. To the present day, raptors have served as symbols for a wide range of human beliefs.

The connection people have to birds of prey led to the keeping of raptors in captivity. This activity reached its zenith with ancient falconers in medieval Europe and the Arab world. Philosophies, traditions, and management practices used in the sport of falconry have been passed down through generations and are today infused with modern technology, science, and medicine. They form the foundation of captive raptor management.

In the modern day, the use of live raptors to enhance environmental education has become increasingly popular in zoological gardens, nature centers, and other educational facilities.

This book is intended as a resource for properly caring for permanently disabled raptors that are used in such educational settings. It is a management tool that guides caretakers through the process of selecting an appropriate bird and then providing it with a high quality of life. It is geared toward the novice raptor handler, but also contains useful information for experienced caretakers and any permitted individuals that have a raptor in their possession.

However, it is NOT intended as a guideline for managing raptors kept for the sport of falconry, captive breeding programs, or free-flight demonstrations. Birds held for these purposes often have different needs, especially in relation to housing and training. It also does not focus on managing captive-bred raptors. Birds bred in captivity often exhibit a variety of different behaviors, based in part on how they were raised. Wild birds have unique spirits that evolve from their experiences in the wild. If they become permanently disabled, they must learn to adapt not only to a captive life but also to the handicap that claimed their freedom.

The contents of this book reflect the experiences of TRC (The Raptor Center, College of Veterinary Medicine, at the University of Minnesota), over the past thirty years in managing permanently disabled raptors, as well as experiences of friends and colleagues. The contents are based on an ethical code of conduct TRC established for maintaining these birds in captivity and reflect the Wildlife Educator's Code of Ethics (page 18) set forth by the National Wildlife Rehabilitator's Association in 2004.

TRC rehabilitates over 700 raptors each year and maintains a collection of thirty permanently disabled birds that work as educational ambassadors. A few of the birds are in static displays only, but the majority is trained to perch on a gloved hand for program use. TRC's standards for caring for and displaying these birds are high. Only comfortable, well-adjusted, healthy birds are partners in TRC's education programs and they are provided the highest quality of life possible in captivity.

As raptor caretakers will attest, successfully managing raptors in captivity is as much an art as a science. Many raptor handlers typically follow the same general methods and would feel at home with the techniques presented here. However, every bird and situation is different and there are a variety of ways to approach any problem. The most successful raptor handlers and caregivers are those who take the time to study these birds in the wild, talk with other handlers, respond to the individual needs of each bird, and forever hold the greatest respect for the spirit of the wild raptor in their care.

Acknowledgements

TRC wants to extend a special thank you to all the people who contributed knowledge and support in the writing of this book.

First, to Dr. Pat Redig, Cofounder and Director of TRC, whose lifelong passion for these magnificent birds created an organization devoted to their diverse needs.

To Mark Martell, coauthor of the first edition.

To TRC's Care and Management Team consisting of, Jane Goggin, Kate Hanson, Lisa Koch, and Jen Vieth.

To Ron Winch for sharing his photographic skills; to Gail Buhl for her photographic skill and artistic talent; to John Karger, Last Chance Forever, for sharing his wealth of experience and innovative equipment designs.

To all TRC staff and volunteers, past and present, who shared ideas, encouragement, and many long days and nights caring for raptors; and to Daisy Ritter for nurturing the original idea.

In addition, a special thank you is extended to many other wildlife educators who provided valuable input in the creation of this book: Marlys Bulander, USFWS Region 3; Recee Collins, USFWS Region 4; Kate Davis, Raptors of the Rockies; Debbie Farley, Spring Brook Nature Center; Dianna Flynt, Florida Audubon; Diana Granadas, Native Bird Connections; Steve Martin, Natural Encounters; Jeff Meshach, World Bird Sanctuary; Mike Pratt, Vermont Raptor Center; Mona Rutger, Back to the Wild; and Len Soucy, The Raptor Trust.

One final thank you is extended to John Arent, for sharing his innate insight of raptor behavior and enlightening many people how to connect with their wild spirits.

A Wildlife Educator's
Code of Ethics

(Reprinted with permission of the
National Wildlife Rehabilitator's Association)

1. A wildlife educator should strive to achieve high standards of animal care and programming through knowledge and training.

2. A wildlife educator should acknowledge limitations and enlist the assistance of a veterinarian or other trained professionals when appropriate.

3. A wildlife educator should respect other educators and persons in related fields, sharing skills and knowledge in the spirit of cooperation for the welfare of the animals.

4. The physical and mental well being of each animal should be a primary consideration in management and presentation.

5. A wildlife educator should strive to provide professional and humane care for the animals in their care, respecting the wildness and maintaining the dignity of each animal in life and in death.

6. Non-releasable animals, which are inappropriate for education, foster parenting, or captive breeding have a right to euthanasia.

7. A wildlife educator must abide by local, state, provincial, and federal laws concerning wildlife and associated activities. Animals must be legally acquired with proper documentation. Animals must go to legal and reputable facilities or individuals.

8. A wildlife educator should establish safe work habits and conditions, abiding by current health and safety practices at all times.

9. A wildlife educator should encourage community support and involvement through public education. The common goal should be to promote a responsible concern for living beings and the welfare of the environment.

10. A wildlife educator should work on the basis of sound ecological principles, incorporating appropriate conservation ethics and an attitude of stewardship.

11. A wildlife educator should conduct all business and activities in a professional manner, with honesty, integrity, compassion, and commitment; realizing that an individual's conduct reflects on the entire field of wildlife and environmental education.

Chapter 1: PERMITS

In the United States, wildlife is considered the property of all citizens and is protected and managed by the federal and state governments. Public sentiment, as well as law, does not favor the unrestricted use of wildlife for commercial or for-profit purposes. Thus killing, collecting, or taking into captivity most forms of wildlife is heavily regulated.

1.1 FEDERAL PERMITS

All birds native to North America (thus excluding pigeons, European starlings, and English sparrows) are protected by at least one, and sometimes up to three, federal laws. Additionally, many states and municipalities also regulate the keeping of wild birds.

1.1a Federal Laws

One of the earliest laws passed to protect wildlife in the United States was the Migratory Bird Treaty Act (MBT), 1918. This law was initially an international treaty between the United States, Canada, Great Britain, and Mexico, and has now been amended to include Russia and Japan. It prohibits anyone from taking, killing, or keeping any native bird, its parts, or its nest, without a permit or license. All raptors native to the United States are covered by this law.

Congress passed the Bald Eagle Act (BEA), 1940, in response to the slaughter of eagles during the first half of the twentieth century, and because of the special status the bald eagle holds as our national symbol. This law, which protects both bald eagles and golden eagles and their nests and nest trees, specifically prohibits anyone from killing or disturbing either species.

The Endangered Species Act (ESA), 1973, provides additional protection for any animal listed as "threatened" or "endangered." The species of raptors included on this federal list change, so make sure to check with your local United States Fish and Wildlife Service (USFWS) office, or its website, www.fws.gov, for the most up-to-date information.

1.1b Permits

Each of these laws has a separate set of regulations and permits. Which permit(s) you will need depends on the species you want to acquire. For example, to possess a red-tailed hawk, you need a Special Purpose Possession Education Permit to keep the bird under the Migratory Bird Treaty Act. To possess a bald eagle, you need an Eagle Exhibition Permit.

Federal permits are issued through the USFWS at their regional offices (appendix A). Each permit requires annual reports and renewal. Zoos and other organizations that are members of the American Zoological Association (AZA) may not need to get a permit for display of non-endangered species. However, TRC would still advise you to check with your regional USFWS office before obtaining any birds.

Federal and state agencies and personnel may also need to acquire permits. This includes state and national parks, wildlife areas, and research facilities.

Generally, non-native raptors (those not regularly found in North America) are not protected under these laws, with a few exceptions. These include rare U.S. visitors, non-native species that look like native species (such as the White-tailed sea eagle) and some hybrids. When two species are hybridized and one is protected under the MBTA, such as a Lanner falcon/Peregrine falcon cross, then the bird is protected under the MBTA.

Even though most non-native species are not protected under the laws listed, there are special regulations governing their import. All raptors are listed under the Convention on International Trade in Endangered Species (CITES). CITES requires special permits from the country of origin, as well as the United States, before a raptor can be brought into this country. Also, the Wild Bird Conservation Act (WBCA) regulates the import of birds into the United States. Again, check with your local USFWS office for specific regulations and permit applications.

1.2 STATE PERMITS

Most states also require both individuals and organizations to obtain permits to keep raptors. These permits, usually issued by a state's natural resource agency (appendix A), are separate from your federal permit and can be more restrictive. For example, some states do not allow permanently disabled birds to be used for education. While state and federal wildlife officials usually work in conjunction with each other, the issuance of one permit does not guarantee the issuance of others.

Some local governments also have regulations or ordinances on keeping wild animals that might affect your ability to house or possess a raptor. It's wise to find out whether your municipality has any restrictions that might affect you.

1.3 PERMIT LIMITATIONS

All of the permits mentioned so far only allow you to possess a bird. If you plan to take a bird to another location (off-site) for programs, you'll need to have specific permission for that on your permit — particularly if you plan to cross state lines. You must have both federal and state approval to move a bird between or through states. This includes both the state from which you are taking the bird and the state into which you are taking it. Having a permit to keep a bird in one state does not give you the right to take it into another.

Don't assume that the possession of a falconry, collection, rehabilitation, or bird-banding permit allows you to keep live birds for educational use. All of these permits are separate; they do not overlap. Similarly, an education permit does not allow you to do rehabilitation, falconry, or bird banding, or to keep dead birds or their parts.

1.4 OTHER CONSIDERATIONS

Some institutions, such as state and federal agencies, universities, and private colleges, are required to follow animal care regulations stipulated by the Institutional Animal Care and Usage Committee (IACUC), a United States Department of Agriculture (USDA) program. The possession of raptors for educational purposes falls under the committee's jurisdiction. This means that the birds must be listed under a management protocol accepted by the committee, and monthly inspections by a committee member may occur.

1.5 RECORD KEEPING

If the regulations and permits governing wild-bird possession seem daunting and complicated, don't feel bad. They are! The paperwork involved in keeping a bird is probably one of the least enjoyable aspects of working with raptors — but it's also one of the most important. It's your responsibility to have the necessary permits, keep them in your possession, and file the required reports in a prompt manner. Failure to do any of these things can result in restrictions, the suspension or revocation of your permit, the loss of your birds, fines, or in extreme cases, prison terms.

Here are a few suggestions for dealing with this paperwork:

- Remember that permit offices are often understaffed and require some lead-time for receiving applications and processing and issuing permits. Give yourself plenty of time (one to two months) to get your permit.

- Have one person in your organization be responsible for handling permits. This person should be responsible for knowing the law, applying for permits, and filing reports.

- Don't throw anything away! Start a file for all of your old permits, reports, and correspondence. Make it a game and see how big a file you can amass. It might seem like a waste of space, but you never know when you'll need to show that you were following regulations when you took a particular action.

- Get to know your local game warden or conservation officer. Take the time to let him or her get to know you and understand that you're interested in both protecting wildlife and following the law. This will make getting a permit easier for you, and will make the officer's job easier as well.

1.6 PERMIT DISPLAY

A copy of your state and federal permits should be located close to your bird housing facilities. This will let people know that the birds are legally possessed and that you are working under the support of the state and federal wildlife agencies. Also, a copy of your permits should accompany the birds whenever they travel off-site. This will help avoid a potentially sticky situation should you get stopped or questioned during the transport of your avian educators.

1.7 LIABILITY/INSURANCE

Although it isn't strictly a permit issue, TRC wants to mention another important legal consideration — liability. Because raptors are capable of injuring people (handlers as well as audience members) you must be prepared for the possibility that your bird will hurt someone. Injuries can lead to lawsuits, and TRC strongly recommends that anyone keeping raptors carry appropriate personal medical and liability insurance. If you are part of an established organization, check with your director or legal staff to ensure you are covered.

1.8 SUMMARY

To possess a live raptor for educational use, you need both federal and state permits. Along with these permits come reporting and record-keeping responsibilities. Contact your local USFWS and state natural resource agency office to obtain the necessary forms. Appendix A contains the names, addresses, and phone numbers of the appropriate contacts for your state.

Chapter 2: SELECTING A BIRD FOR EDUCATION

There are fifty-four species of raptors (birds of prey) in North America, and more than 400 species worldwide. They exhibit a wide diversity of natural history, size, coloration, and behavior. When selecting a bird to use in your project, you need to consider many factors: the legalities of possessing a particular species (see chapter 1, Permits), the bird's intended use, management issues, the experience of the handlers, and the environmental message you wish to convey. You no doubt will be excited about the new adventure you are preparing to embark on, but start slow. Do your homework in preparation for a single bird and then manage it successfully before considering additions to your collection.

This chapter offers some generalizations, guidelines, and recommendations for choosing a bird based on TRC's experience and discussions with others. The danger of making such generalizations is that someone, somewhere, will have successfully accomplished something that is not recommended in this chapter. It is important to realize that handlers, as well as birds, are individuals, and circumstances might allow some birds to be used in situations TRC does not recommend here. What follows then, are guidelines for people inexperienced with the wide range of North American raptor species and their behavior. These guidelines are written for display and program birds, not demonstration birds (see definitions below).

2.1 IMPORTANT CONSIDERATIONS

2.1a Legalities

As discussed in chapter 1, Permits, possession of a raptor requires permits from federal and state authorities. Therefore, one of the first things you'll need to know is whether you can get a permit for the particular species you are interested in obtaining. It's advisable to consult with your local USFWS office and state natural resource agency early in the planning stage, particularly for less common species.

2.1b Display Versus Program Use

One of your first considerations is deciding how your bird is going to be used. Will it be a display bird, to be placed in an exhibit with little or no handling, a program bird, to be trained (manned) and displayed on a gloved hand, or a demonstration bird, to be flown in front of people?

A program bird will need more training and handling than a display bird. Thus, larger or more difficult-to-handle birds might be better suited for display use than for programs. Some individual birds are less amenable to handling or changes in their management, and can be comfortably put on display but not taken out on programs.

Can birds be used for both program and display? The answer is a qualified yes; some birds can be housed in a display and still be trained to step up on a glove for use in a program. Other birds, however, resist this kind of management. Many raptors, especially eagles and falcons, resort to their wildness when free-lofted in a display and might not be good birds for program use unless tethered.

Only raptor experts should embark in developing a free-flight demonstration program. The training and management of fully flighted raptors is another book in itself and to do it safely for everyone concerned requires a strong knowledge of bird behavior and weight management. If you feel you are interested and ready to embark on such a journey, contact your state and federal permitting agencies to find out about possible restrictions and make sure you have liability insurance. TRC also strongly recommends that you apprentice under an experienced handler or organization that has a reputable history of delivering safe free-flight raptor programs.

Large organizations experienced in free-flight programs include the following. A description of these organizations along with contact information can be found on the web.

International Association of Avian Trainers and Educators (IAATE)
www.iaate.org

Last Chance Forever (LCF), San Antonio, Texas
www.lastchanceforever.org

Natural Encounters Inc. (NEI)
www.naturalencounters.com

National Bird of Prey Centre, Gloucester, England
www.nbpc.co.uk

World Bird Sanctuary (WBS), St. Louis, MO
www.worldbirdsanctuary.org

Keep in mind that people don't have to see raptors fly to become enthralled and understand your program's key messages. Well-managed, healthy birds that are solidly trained to display calmly on a gloved hand or on a perch are just as awe-inspiring and effective.

2.1c Management Issues

When selecting a species for your education program, a major consideration is management. Will you be able to meet the bird's needs and keep it comfortable and healthy during its stay with you? Here are some specific management issues to think about:

Life span
Raptors can live a long time in captivity if they are managed properly. Smaller species have shorter life spans than larger birds. For example, an American kestrel can live around fifteen years in captivity, a red-tailed hawk thirty years and a bald eagle fifty years. Thus, when agreeing to accept a bird into your program, you are making a long commitment for the life of the bird.

Before making such an important decision, it is recommended to form a team of people to evaluate both your desire to have a bird in the first place and the realities of potentially caring for a bird for decades. You need to think hard about the history and long-term goals for your program. Is your program established and expected to be active in fifteen or thirty or fifty years? Do you have frequent staff turnover that would often result in new caretakers and trainers? Every bird responds differently to different people and new staff will require a period of training themselves with the previous caretakers and trainers. If adequate training of new staff is not conducted, a well-trained bird can become unmanageable. Written guidelines for bird care and training protocols are a must to help make smooth transitions between caretakers and maintain a level of consistent care.

Cost
Maintaining a raptor in captivity is not free. Even after the initial expenses of building suitable housing, purchasing equipment (scale, handling equipment, etc.) and possibly paying for airline travel to get the bird to you, the cost of captivity continues. The bird will need a constant supply of high quality food, to have its housing facility maintained, and to have appropriate medical and management care. Beak and nail trims, fecal tests, vaccinations, annual

medical check-ups, and veterinary care during times of illness or injury all cost money.

Make sure you have a budget secured to take care of your feathered educator before selecting and acquiring a bird. Contact your veterinarian to get a cost estimate for basic and emergency services. Remember, as a raptor caretaker, you are responsible for the daily quality of life and health of your bird for its entire life. This is a big responsibility and can be costly.

Staffing

Another management issue to consider is your staffing. One or two people should be designated as the leader in the care and management of your bird. The experience level of this person will dictate what type of bird is most suitable for you. Some species are more challenging to manage and should only be handled/trained by people with extensive raptor experience. Bald and golden eagles fall into this category. They are not "beginner's birds."

In the following pages, TRC has rated different species based on their level of difficulty for managing and training. Also listed are the top six species TRC considers to be solid, manageable birds that are often successfully managed by novice handlers. However, TRC strongly believes that before acquiring any species, a novice raptor caretaker should go through training with an experienced raptor manager.

One other staff issue to think about is time. Maintaining a raptor takes time every day. More time will be needed for a program or demonstration bird, especially during the initial training phase. However, even after a bird is trained or has attained a level of comfort in its display, it will need care 365 days a year. Feeding, cleaning, weighing, weathering, and continued training are all activities that must be accomplished. Make sure that you can commit the necessary time now and in the future.

2.1d Bird Specifics

The section that follows will help you decide on an appropriate species that is compatible with your experience level, space, and program goals. Once you have made the commitment to acquire a certain species and have all the preliminary work done, it is time to find the individual bird that is right for you. People acquire raptors from one of three sources: rehabilitation facilities that are looking to place permanently disabled birds, licensed captive breeding programs that sell them, and education facilities that surplus animals for various reasons. This book is written primarily for wild birds acquired from rehabilitation facilities.

Before agreeing to accept any bird, find out as much information about it as possible. Keep in mind that if an educator or education facility is looking to place a bird elsewhere there must be reason. Find out what this reason is. If the reason revolves around the fact that the bird has an undesirable behavior or health issue, make sure you can and are willing to deal with it. Moving a bird to a new location or situation does not mean that the problem will disappear.

In North America, many of the raptors used in education programs are received as permanently disabled birds from a rehabilitation facility. Once again, find out as much information as possible regarding the bird in question before agreeing to take it. TRC recommends finding out the date of admission (how long it has been in captivity), admission cause, injury and its status, how the bird has been housed, its age, and its temperament. Also, it is a good idea to get a few pictures of the bird to see its overall condition.

Some birds, due to their injuries and/or temperament, should not be kept in captivity (see below). They are prone to further injury, do not train well, and are generally miserable if confined. Quality of life must be a major consideration when accepting a raptor into your education program. Do not depend on the organization looking to place a bird to make that decision. Everyone has his or her own idea of quality of life and it is much better for you to make an educated decision before getting a bird. If you are not sure whether or not to accept a bird, contact other educators or raptor professionals to get other opinions. You should also check with your state wildlife licensing office to see if any restrictions exist in your state regarding a bird's injury.

TRC has high standards for placing permanently disabled birds. Here are some of the qualities looked for in a placement candidate:

1. A raptor must tolerate captivity well (eat regularly, accept its housing without constantly bouncing off the walls and/or injuring itself).

2. A raptor must maintain good feather, foot, and wrist condition.

3. A raptor must not have an injury that:

   • Compromises balance (such as full-wing amputations, permanent wing nerve damage, poorly healed leg or pelvic fractures, amputated leg or foot)

   • Permanently alters behavior (such as head trauma resulting in uncontrolled head movements, periodic seizures, or balance problems)

- Has resulted in complete blindness in both eyes

- Or will result in chronic pain (such as arthritis from joint injuries, fractures that result in a non-union of the bone ends, or osteomyelitis of the bone ends).

The latter two qualities can be identified by obtaining a complete medical history of the bird in question, including radiographs of fractures and their status, and pictures of the bird. TRC does not recommend taking any bird that has an active injury, foot problems, or behavioral issues such as aggression. These issues will only turn into nightmares. The first quality, however, is somewhat subjective, but over the years TRC has been able to identify some key factors that help predict a bird's tolerance for captivity.

Age

As a general rule, young animals adapt faster than older animals to new or strange situations. This has important ramifications for selecting a bird. First, it is often difficult to shape the behavior of an older bird that is "set in its wild ways." Falconers refer to adult wild hawks as "haggards." Cantankerous, irritable people, as defined by Webster, are referred to as "old hags," a recognition of the similarity between the two personality types.

Adult red-tailed hawks are a good example. They often act stressed when exposed to people. They keep their mouths open, their tongues sticking out, and when being glove trained, often stand with their hackles and wings raised and their bodies leaning away from the handler. These postures often don't disappear and are an indicator of the bird's chronic discomfort with the situation. A younger bird, while not totally malleable, will adapt more easily to the wide variety of new situations it will encounter.

Another age-related consideration is the effect of the stress of a new environment on a bird's health. Changes in routine that come with a new home may place stress on an older bird, which in turn can lead to reduced appetite or suppressed immune function. All birds require an adjustment period; however, some take a long time, and others never fully adjust. A bird should not be displayed to the public or used in programs until it is comfortable in its surroundings.

Young birds, however, come with their own problems. After a few years they can become so acclimated to their surroundings that they become territorial, or even lay eggs in captivity. This can be a real problem if you try to use these birds for programs — or even approach them for feeding during this time. Aggression toward people can result in injury to the bird and/or to people and make it impossible to use that bird during the reproductive season.

None of this is intended to mean that no adult birds can be acclimated to captivity or that all young birds develop into springtime terrors. However, do keep in mind that the behaviors listed might occur. TRC recommends choosing young birds to train as program birds. Although some adults can be effectively manned and reach a level of comfort, most often their quality of life is higher if housed in a display with minimal handling and areas to hide from the public for relief.

Sex

The gender of a raptor will determine its size, behavior patterns and, in a few species, its color. Most raptors exhibit "reversed sexual dimorphism," which means that females are larger than males. Accipiters and falcons show the most extreme examples of size differences between the sexes. Size difference alone, however, does not greatly affect how the birds are kept in captivity. Males and females have the same basic needs (chapter 4, Housing). The factor most influenced by sex is behavior. TRC has found that the males of many species tend to be a little more high-strung than the females.

Human imprints

Sexual imprinting is a behavioral adaptation that establishes a bird's identity and gives it a fixed visual perception of what its future mate should look like. Birds typically imprint when they are very young. Some waterfowl, for example, imprint only thirteen hours after they hatch. Raptor imprinting seems to take longer (this varies by species) and occurs during the period when the birds' eyes are beginning to focus. In the wild, young birds sexually imprint on their parents. Later in life they will try to mate with — and defend their territory against — something that looks like mom or dad.

Imprinting is a good strategy in the wild. Birds raised in captivity, however, can sexually "imprint" on their human caretakers. These birds cannot be released into the wild because when they reach sexual maturity, they may try to mate with, and/or aggressively defend their territory against, humans. So, these human imprints are often available for placement in educational facilities. Many people desire human-imprinted birds for educational use because when the birds are young, they are usually very "tame" and quite appealing. However, there are several negative aspects associated with human imprints as well.

First, when they reach sexual maturity, they frequently become territorial and aggressive toward people during the breeding season, can be difficult to handle, and are potentially dangerous to their handlers. TRC strongly recommends keeping these birds tethered during the breeding season (see chapter 8, Training). Second, human imprints can retain food-begging behaviors as adults, which may cause them to "scream," often loudly, when they see their handlers. This can be very distracting during an education program. Lastly, human-imprinted birds are sensitive to management changes and may become destructive or resort to feather plucking behavior if stressed or bored.

Flightless raptors/amputees
People often mistakenly desire completely flightless raptors for display or program use, with the thought that they are easier to manage since their movement is greatly restricted. This very fact causes them to be more difficult to manage. Most often, raptors become completely flightless due to full wing amputations or permanent wing nerve damage. Although their space requirements are often less than those of fully or partially flighted birds, they have unique management issues.

First, they have a hard time maintaining their balance as they move, and if they topple over, often cannot right themselves. Second, completely flightless birds, and full-wing amputees in particular, are prone to developing sores on the bottoms of their feet (chapter 7, Medical Care). They are off balance due to the weight difference of each side, resulting from missing a wing and/or the extreme pectoral muscle atrophy on the affected side. Third, birds missing a wing or lacking use of a wing due to nerve damage can easily drown if their bath water is too deep. Fourth, birds with nerve damage may grab their damaged wing when trying to recover balance and cause soft tissue damage, step on their wing and fall over, or self mutilate their wing if nerve sensation is still present.

In the case of full wing amputees, people often think that a bird with such a disability will alter its behavior to adjust to its handicap. It has been TRC's experience that this is not the case and even after ten years, amputees think they can fly and will launch themselves off a perch, injuring their amputation site or falling on their side unable to get up. TRC believes that the quality of life for these birds is very low and does not recommend keeping them. Sufficient numbers of full-winged, permanently disabled birds are available for placement.

2.1e Program Goals

It is critical that you give a lot of thought to how and why you are using a raptor. The how part — where, when, and by whom — are the issues addressed in this book. The why is something not covered here, but it must be considered. It involves the message you are trying to deliver to your audience and the way in which you want to deliver it.

For example, are you trying to show a sample of your local natural history, or present a global view? Is your message about common backyard species, or endangered and threatened animals? The species of raptor you choose should reflect your intended message. Develop your education program and its goals first and then choose a raptor that will enhance the delivery of your message.

2.2 SPECIES ACCOUNTS

In the species accounts that follow, TRC summarizes the pluses and minuses of thirty-five of the most commonly used raptor species. If you are interested in obtaining a species not mentioned here, refer to natural history references and the Internet for species-specific information (diet, weight range, nesting behavior, habitat, etc.). You can also contact other educators, rehabilitators, or wildlife biologists in your area that might be able to provide information on the captive behavior and diet of the species in question.

The species accounts presented here reflect the experiences of TRC as well as those of other experienced raptor managers across the country that have cared for a wide range of species. It is important to know that these recommendations are based on the following:

- The birds were previously wild, permanently disabled birds acquired from rehabilitation facilities, not birds acquired from captive breeding programs.

- The birds trained for program use are young (less than one year old) not adults.

- The information listed for display birds pertains to either young or adult birds.

The first six species listed are what TRC refers to as the "steady six." These are species that have proven to be relatively easy for novice handlers to train and manage, and are traditionally solid, consistent educators once trained. The other species are listed in alphabetical order by classification.

The stars after the species name represent TRC's overall rating of the species, from zero to five, with five stars being highly recommended. This rating is based on a species tolerance for captivi-

Fig. 2.1a
Americal Kestrel - male

Fig. 2.1b
American Kestel - female

ty and manning, behavioral characteristics, and overall management considerations. Under the species name are several icons indicating common diet in captivity, recommendations for program and/or display use, experience level required to train and manage the species, the presence of specific medical concerns, and specific temperature considerations (table 2.1).

2.2a The Steady Six

American Kestrel (*Falco sparverius*) ✰ ✰ ✰ ✰ ✰

Description: American kestrels are the smallest North American falcons. The sexes have different color plumage even during their juvenile year. Males have slate gray wings and a rust tail bordered with a dark terminal band. Females have rust/ cinnamon wings streaked with black and a heavily barred cinnamon tail. They are widespread throughout urban and rural North America and hunt small rodents, insects, and sometimes small birds. American kestrels display the long wing structure, dark eyes, malar cheek stripes, and notched beak of the Falco genus.

Management skill level: Novice

Handling and husbandry: American kestrels are readily manned for program use and adapt well to displays. They are cavity nesters and will use a hutch or nest box if provided. Due to their small size, their weights must be monitored closely, especially in cool or variable climates. Also, their notched beak requires expertise when coping to maintain its natural shape. American kestrels bathe regularly.

Temperature: American kestrels need supplemental heat if the temperature in their housing dips below 20°F (–6.7°C).

Eastern Screech Owl (*Megascops asio*) ✰ ✰ ✰ ✰ ✰

Description: Eastern screech owls live in wooded habitats east of the Rocky Mountains in both rural and urban areas. They are small owls with feather tufts, yellow eyes, and a pale beak. They are known for red and gray color phases, and for their wide range of vocalizations including hoots, barks, rasps, screeches, and descending trills. They prey primarily on small rodents and insects and sometimes eat small birds.

Fig. 2.2
Eastern Screech Owl

Management skill level: Novice

Handling and husbandry: Eastern screech owls are easy to man for program use and adapt well to a display. They are cavity nesters and should be provided with a cavity structure/nest box to provide shelter and a place for them to escape stressful situations. They are sensitive to management changes and may go off food if moved to a new location or taken off site for overnight travel. If they wear equipment (anklets and jesses), their feathered legs must be monitored for skin irritation.

Fig. 2.3
Great Horned Owl

Great Horned Owl (*Bubo virginianus*) ☆☆☆☆☆

N

Description: Great horned owls have the most extensive range of all North American owls, adapting well to a variety of habitats. These large, hardy owls are characterized by yellow eyes, a black beak, a distinct facial disk, large feather tufts, and a white throat patch. They have a wide prey base including large and small rodents, other mammals including skunks, and a variety of bird species (including crows and other raptors).

Management skill level: Novice

Handling and husbandry: Great horned owls are solid birds that man well for program use, adapt to traveling, and handle display settings well when properly acclimated. They are early nesters and may exhibit breeding, territorial, or aggressive behaviors starting in early winter. If they wear equipment (anklets and jesses), their feathered legs must be monitored for irritation.

Fig. 2.4
Northern Saw-whet Owl

Fig. 2.5
Red-tailed Hawk

Medical: Great horned owls are very susceptible to West Nile virus and should be housed with appropriate mosquito protection. Vaccination is also recommended.

Northern Saw-whet Owl (*Aegolius acadicus*) ✩✩✩✩✩

 N

Description: Northern saw-whet owls are small, tuftless, forest-dwelling, cavity-nesting owls that inhabit the northern part of the United States, parts of the western United States and Canada. Their adult plumage is grayish brown streaked with white and is spotted on the back. They have a short tail, yellow eyes, a dark beak, white eyebrow and mustache markings, and a light facial disk bordered in brown. Juvenile birds (less than three months old) are dark brown with a buff chest and distinct white eyebrow and mustache markings. Saw-whet owls prey on small rodents, birds, and insects.

Management skill level: Novice

Handling and husbandry: Northern saw-whet owls are very adaptable birds that man well and acclimate to a display. They appear most comfortable if a cavity or nest box is provided in their enclosure, and if two or more saw-whets (not human-imprints) are housed together. When first moved to a new location, they may refuse to eat for a day or two. If they wear equipment (anklets and jesses), their feathered legs must be monitored for irritation.

Medical: Northern saw-whet owls have been reported to contract avian malaria, a mosquito transmitted blood disease, and should be housed with appropriate mosquito protection.

Red-tailed Hawk (*Buteo jamaicensis*) ✩✩✩✩✩

 N

Description: Red-tailed hawks are large buteos found across North America. They have long broad wings and a broad tail that acquires the characteristic red color at about one year of age, after their first molt. They typically inhabit open areas with adjacent trees or nesting structures. Well documented plumage variations within the species add interesting regional differences. They prey on a variety of animals including mice, rats, rabbits, snakes, lizards, and pheasants.

Management skill level: Novice

Handling and husbandry: Well known for adapting readily to display and program situations, red-tailed hawks are often the most reliable birds in a collection. They are considered relatively easy to train and manage, and they often develop calm demeanors. They adapt well to people, travel, and new environments.

Medical: Young red-tailed hawks are susceptible to aspergillosis when first brought into captivity (especially in the Midwest and humid states). To be safe, it is recommended to put them on a preventative course of treatment when they are first acquired, undergo a major management change, or become ill. They are also highly susceptible to West Nile virus and should be housed with appropriate mosquito protection. Vaccination is recommended.

Western Screech Owl (*Otus kennicottii*) ☆☆☆ ☆☆

 N

Description: Western screech owls are small tufted owls that inhabit riverside woodland and mesquite scrub, and the cactus desert west of the Rocky Mountains. The pale gray color morphology is most prevalent in the dry southwest with a darker, browner morphology prevalent in the humid northwest. The red color phase is rare but occurs in the Pacific Northwest. Their diet mainly consists of insects, such as beetles, moths, and worms; small rodents such as mice, shrews, and kangaroo rats, small birds, frogs, and salamanders. They often hunt on the wing.

Management skill level: Novice

Handling and husbandry: Western screech owls are easy to man for program use and adapt well to a display. They are cavity nesters and are most comfortable with a nest or shelter box in their enclosure. If they wear equipment (anklets and jesses), their feathered legs must be monitored for irritation.

Fig. 2.6
Western Screech Owl

Fig. 2.7
Cooper's Hawk

2.2b Accipiters

Cooper's Hawk (*Accipiter cooperi*)

 E

Description: Cooper's hawks are medium-sized accipiters (bird-eating hawks) that inhabit forested areas in rural and urban settings. Adults have blue/gray wings, head and back, and orange-to-red eyes. They have the typical disposition of accipiters: hyperactive, nervous, and self-destructive in captivity.

Management skill level: Expert

Handling and husbandry: Due to their hyperactive, nervous disposition they are not recommended for program or display use. They do not man well and in a display (especially a small one) panic and often injure themselves as they hit the walls, ceiling, etc.

Medical: Captive Cooper's hawks commonly exhibit hyperactivity which often results in soft tissue damage to their cere, eye ridges, and head, and broken wing and tail feathers. They are also highly susceptible to West Nile virus and should be housed with appropriate mosquito protection. Vaccination is recommended.

Temperature: Most Cooper's hawks migrate from cold climates and may need an additional heat source if the temperature in their housing dips below 20°F (–6.7°C).

Fig. 2.8
Northern Goshawk

Northern Goshawk (*Accipiter gentilis*) ☆

 E

Description: Northern goshawks are the largest of the North American accipiters. They inhabit deep forested areas and are generally quite secretive. Adults have the characteristic accipiter color: blue/gray wings, back and head, and orange-to-red eyes. They are also characterized by white eyebrows. Their diet consists of birds and mammals, such as red squirrels and snowshoe hare. They are typically high-strung and nervous at any age.

Management skill level: Expert

Handling and husbandry: Due to their hyperactive behavior in captivity, they are not recommended for program or display use. They do not man easily and act constantly stressed by close proximity to large groups of humans. They are trained and used in the sport of falconry, but their energy and aggression are targeted at the natural behavior of hunting. Care must be taken when managing them to prevent breakage of their long tail feathers, and cere and wrist damage if they are free-lofted.

Medical: Northern goshawks are highly susceptible to aspergillosis in captivity. They should be put on a preventative course of treatment when first acquired and anytime a major management change occurs. They are also highly susceptible to West Nile virus and should be housed with appropriate mosquito protection. Vaccination is recommended.

Sharp-shinned hawk (*Accipiter striatus*)

 E

Description: Sharp-shinned hawks are the smallest North American accipiter. They inhabit boreal forests in the northern part of the country and Canada and prey primarily on small birds. Young birds have plumage typical of juvenile accipiters: a brownish back and tail and yellow eyes. Adults are blue/gray with orange-to-red eyes. They are typically nervous, skittish birds in captivity.

Management skill level: Expert

Handling and husbandry: Due to their nervous, hyperactive disposition when confined, they are not recommended for program or display use. They are easily stressed and do not tolerate exposure to people well. If free-lofted, they must be monitored closely for cere and wrist damage.

Temperature: Most sharp-shinned hawks migrate from cold climates and would need an additional heat source if the temperature in their housing dips below 20°F (−6.7°C).

Fig. 2.9
Sharp-shinned Hawk

Fig. 2.10
Broad-winged Hawk

2.2c Buteos

Broad-winged Hawk (*Buteo platypterus*) ☆ ☆ ☆

Description: Broad-winged hawks are small brown buteos that inhabit forested areas in the eastern half of the United States. Adults have a horizontally barred chest, dark chocolate eyes, and a tail banded with distinct thick bands alternating light/dark. They are migratory, forming large kettles on their journey to Central and South America and eat primarily small rodents, amphibians, and snakes.

Management skill level: Intermediate

Handling and husbandry: Broad-winged hawks can be used for program or display use, although they are more high-strung than red-tailed hawks. They are prone to excessive weight gain and their diet must be managed closely.

Temperature: Broad-winged hawks do not tolerate cold temperatures well and must be provided with an additional heat source if the temperature in their housing dips below 20°F (–6.7°C).

Fig. 2.11
Ferruginous Hawk

Ferruginous Hawk (*Buteo regalis*) ☆

Description: Ferruginous hawks are the largest North American buteos and are characterized by long tapered wings with dark tips, dark leggings which look like a "V" on the lower abdomen, and a large yellow gape. They inhabit grasslands in the western United States and feed primarily on small mammals such as rabbits and prairie dogs. Sometimes they will feed on a variety of other prey items such as birds, reptiles, amphibians, and insects.

Management skill level: Expert

Handling and husbandry: Ferruginous hawks often act like small booted eagles. Their size, power and aggressive tendencies can make them dangerous for program use. However, they can be managed in a display and young birds can be manned for program use by an experienced handler. If they wear equipment (anklets and jesses), their feathered legs must be monitored for irritation.

Medical: Ferruginous hawks are susceptible to aspergillosis (especially if brought into a northern or humid climate) and should be treated with a preventative drug when first acquired, undergo a major management change, or become ill. They are also prone to obesity and bumblefoot in captivity.

Fig. 2.12
Red-shouldered Hawk

Red-shouldered Hawk (*Buteo lineatus*) ☆ ☆ ☆

Description: Red-shouldered hawks are medium-sized buteos that inhabit damp woodland areas usually surrounding a body of water. Adults possess distinct rust wing patches that give them their name, a strikingly barred tail, and a cinnamon chest. Their range includes the western-most edge of Oregon and California, and the eastern half of the United States. They prey on small rodents, birds, amphibians, and snakes. They are relatively high-strung birds that are very vocal during the breeding season.

Management skill level: Intermediate

Handling and husbandry: Red-shouldered hawks are more excitable than some of the other buteos. Young birds can be trained to the glove and all ages can be acclimated to a display provided the area is large enough to give them room to retreat from close proximity to people. In a small space they often hit the walls in panic and damage their cere, wrists, or feathers. Red-shouldered hawks have long tails and care must be taken when housing them to prevent tail feather breakage.

Fig. 2.13
Rough-legged Hawk

Temperature: The northern birds are migratory and should be monitored closely in extremely cold climates for signs of cold stress.

Rough-legged hawk (*Buteo lagopus*) ☆ ☆ ☆

Description: Rough-legged hawks are large northern buteos characterized by feathers that extend down to their feet and a light colored tail with a wide dark terminal band. They come in several different plumage variations (including light and dark morphologies). The most common plumage is characterized by a "comma" of dark feathers on the underside of the wings at the wrist and a dark belly band. Rough-legged hawks eat primarily small rodents.

Management skill level: Intermediate

Handling and husbandry: Rough-legged hawks can be effective for both program and display use. If they wear equipment (anklets and jesses), their feathered legs must be monitored for skin irritation. They also are prone to excessive weight gain and their diet must be monitored closely.

Medical: Rough-legged hawks are very susceptible to bumblefoot in captivity. Managing their weight to prevent obesity and providing them with different sizes and textures of perches will help keep their feet healthy. They are also extremely susceptible to aspergillosis when stressed and should be treated with a preventative drug when acquired, undergo a major change in management, or become ill.

Temperature: Rough-legged hawks are northern climate birds that do not tolerate warm temperatures well.

Swainson's Hawk (*Buteo swainsoni*) ☆ ☆

Description: Swainson's hawks are long-winged large buteos that inhabit open areas in the western half of the United States and Southern Canada. Adults are easily identified by a bib of dark feathers contrasting with a lighter chest. They migrate to the pampas of southern Argentina in the winter and prey on small mammals and insects.

Management skill level: Advanced

Handling and husbandry: Swainson's hawks have a reputation of being actively aggressive towards people. For this reason they are recommended for program use only by experienced trainers and handlers. They can also be managed in a display.

Fig. 2.14
Swainson's Hawk

Temperature: These birds are migratory and should be monitored closely. They may need an additional heat source if the temperature in their housing dips below 20°F (–6.7°C).

2.2d Caracara

Crested Caracara (*Polyborus plancus*)

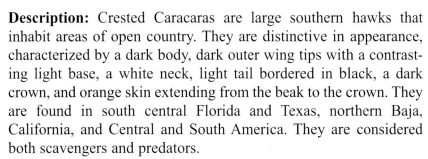

Description: Crested Caracaras are large southern hawks that inhabit areas of open country. They are distinctive in appearance, characterized by a dark body, dark outer wing tips with a contrasting light base, a white neck, light tail bordered in black, a dark crown, and orange skin extending from the beak to the crown. They are found in south central Florida and Texas, northern Baja, California, and Central and South America. They are considered both scavengers and predators.

Management skill level: Expert

Handling and husbandry: Young caracaras can be manned for program use, although they are vocal when handled, bite hard, and are known for plucking their feathers. They are active, highly intelligent birds that do well in a display but require environmental enrichment to avoid destructive behaviors. They are proficient diggers, so their enclosure must be designed to prevent escape. They are also proficient on the ground and can run fast.

Temperature: Caracaras are warm weather birds and should be provided with supplemental heat if the temperature in their housing dips below 32°F (0°C).

2.2e Eagles

Bald Eagle (*Haliaeetus leucocephalus*)

Description: Bald eagles are fish eagles that inhabit areas near large bodies of water in North America, parts of Central America, and Canada. Birds less than one year of age are mostly dark brown (head and tail included) with a black beak and dark eyes. Their adult "look"

Fig. 2.15
Crested Caracara

Fig. 2.16
Bald Eagle

is not acquired until their fifth year of life and is characterized by a dark brown body, white head and tail, yellow beak, and yellow eyes. In addition to eating fish, they will scavenge for carrion.

Management skill level: Expert

Handling and husbandry: Bald eagles are large, heavy-bodied birds that can be aggressive. Biting, footing, and wing slapping are all common defensive behaviors. Young birds can be manned for program use, but require a person experienced in raptor behavior and handling. Attempting to train adult birds is not recommended, as they are often nervous, high strung, and set in their ways. Both ages can be managed effectively in a display setting, but when free-lofted they must be monitored for damaged wrists and feet. Bald eagles will bathe regularly.

Medical: In a display or transport carrier, bald eagles are prone to developing wrist injuries if spooked or not acclimated/trained properly.

Fig. 2.17
Golden Eagle

Golden Eagle (*Aquila chrysaetos*)

Description: Golden eagles are the only booted eagles found in North America. They occur in many types of habitats, but avoid densely forested areas. They are characterized by a dark brown body, golden cap, feathers extending down to their feet, and in young birds, a white tail with a dark terminal band. They hunt medium-sized mammals including rabbits and hares, game birds, reptiles, and carrion.

Management skill level: Expert

Handling and husbandry: Golden eagles are large, powerful birds that can be trained for both program and display use. Advanced training techniques are critical to curb potential aggressive behaviors when on the gloved hand (such as repeatedly squeezing the gloved hand/arm hard). Golden eagles tend to form bonds with single individuals and are more manageable on the gloved hand if trained and handled by one or a few people. If they wear equipment (anklets and jesses), their feathered legs must be monitored for irritation.

Medical: Golden eagles are prone to bumblefoot in captivity. The condition of their feet must be monitored closely. They are also highly susceptible to aspergillosis, especially when first brought into captivity, major management changes are made, or if they get sick or are otherwise physically stressed. Preventative treatment is highly recommended during these times. In addition, golden eagles are also highly susceptible to West Nile virus and should be housed with adequate mosquito protection. Vaccination is recommended.

2.2f Falcons

Merlin (*Falco columbarius*) ✩ ✩ ✩

Description: Merlins are small falcons that breed in the northern portions of the United States and typically occupy habitats with coniferous forests near bogs, lakes, or forest edges. The females and juveniles of both sexes are generally brown in color while the adult males possess blue-gray feathers on their wings and back. They eat primarily small birds.

Fig. 2.18
Merlin

Management skill level: Advanced

Handling and husbandry: Merlins are energetic little falcons that can be manned for program use and acclimated to a display. They are more high-strung than American kestrels and if free-lofted are prone to breaking feathers if not managed properly. Many individuals become vocal, which can be a distraction during education programs. Merlins will bathe regularly.

Medical: Many individuals have been reported to self mutilate their wings or legs in response to highly stressful situations. They are also highly susceptible to trichomoniasis and avian malaria. Thus, a wild bird diet is not recommended and mosquito-proof housing is essential.

Temperature: Merlins require an additional heat source if the temperature in their housing dips below 20°F (−6.7°C).

Fig. 2.19
Peregrine Falcon

Peregrine Falcon (*Falco peregrinus*) ☆☆☆

Description: Peregrine falcons are medium/large falcons that inhabit many continents and feed primarily on bird species. They are fast aerial creatures that came off the federal endangered species list in 1999. Juvenile birds are generally brown in color, with darker vertical streaks on their chest and darker bars on their feathers. Adult birds acquire a bluish/gray color on their wings, back, and head, and maintain a lighter chest with varying degrees of horizontal barring. Both ages have distinct malar stripes.

Management skill level: Advanced

Handling and husbandry: Peregrine falcons can be manned for program use and handle a display setting well if properly acclimated. Females tend to be more relaxed than males in captivity. Their notched beak requires expertise when coping. Peregrine falcons bathe regularly.

Medical: Peregrine falcons are very susceptible to developing bumblefoot if kept over-weight, their talons get too long, and/or they are not provided with adequate flat surfaces to perch on. They also require a bird diet to maintain their health.

Temperature: Many peregrines migrate out of cold climates and may require an additional heat source if the temperature in their housing dips below 20°F (−6.7°C).

Prairie Falcons (*Falco mexicanus*) ☆☆

Description: Prairie falcons are medium/large falcons that inhabit open, dry habitats in the west and southwestern states. Both juveniles and adults are brownish in color with a lighter streaked chest and thin brown malar stripes. They hunt a variety of mammal and bird species, including ground squirrels.

Management skill level: Advanced

Fig. 2.20
Prarie Falcon

Handling and husbandry: Prairie falcons are typically high-strung, active, vocal falcons that with extreme patience can be manned and acclimated to a display situation. Trained birds with some flight ability are often better managed if tethered. Their notched beak requires expertise when coping.

Medical: Wild prairie falcons often get a parasitic infection of nematodes called serratospiculum (air sac worms) that should be treated. They also have been known to develop an untreatable "star gazing" neurological disorder, the cause of which is unknown.

Temperature: Prairie falcons can develop frostbite on their feet in cold temperatures and may require an additional heat source if the temperature in their housing dips below 20°F (–6.7°C).

2.2g Harriers

Northern Harrier (*Circus cyaneus*) ☆ ☆

Fig. 2.21a
Northern Harrier - male

Description: There are ten species of harriers worldwide, but only Northern harriers are found in North America. They are ground nesting medium-sized hawks that inhabit grasslands and marshes and hunt with a "quartering style" of flight. They have long wings, a long tail, and adults differ in color. Females are generally brown with barred wings, a barred tail, a buff streaked chest, and pale yellow eyes. Adult males have a light blue/gray back and head, dark outer wing tips, a dark wing edge that contrasts with light inner flight feathers, and a lemon colored eye. Juvenile birds of both sexes resemble the female in color. Northern harriers have large ear openings and their keen sense of hearing aids in the capture of their prey, primarily small rodents.

Fig. 2.21b
Northern Harrier - female

Management skill level: Advanced

Handling and husbandry: Northern harriers are skittish, high-strung birds that can be manned for program use with patience and time. They handle a display setting as long as ample room is provided. Care must be taken when handling and housing harriers to prevent their long tail feathers from breaking.

Fig. 2.22
Mississippi Kite

Temperature: Northern harriers do not tolerate cold temperatures well and require a heat source if the temperature in their housing dips below 32°F (0°C).

2.2h Kites

Mississippi Kite (*Ictinia mississippiensis*) ☆☆☆☆

Description: Mississippi kites are medium-sized birds with a gray body, narrow pointed wings, a long dark tail, and black mask around the eyes. They inhabit open areas dissected by brush and low trees. They primarily hunt and eat large insects on the wing, but will also prey on small bats, mammals, amphibians, and reptiles.

Management skill level: Intermediate

Handling and husbandry: Mississippi kites are active birds with a gentle demeanor. Young kites man well and both glove trained and display birds do best when housed with other Mississippi Kites or Swallow-tailed kites. They have soft legs with short shanks and equipment must be monitored carefully. Also, their long tails are prone to breakage without proper housing. They sometimes can be difficult to get to eat in captivity.

Temperature: Mississippi kites are warm weather birds that require a heat source if the temperature in their housing dips below 32° F (0° C).

Swallow-tailed Kite (*Elanoides forficatus*) ☆☆☆☆

Description: Swallow-tailed kites are medium-sized birds with a long black deeply forked tail, dark wing tips and a white body and head. They inhabit lowland cypress swamps or tall riverine woodlands in the eastern edge of South Carolina, Florida, the southernmost edge of Louisiana, Mexico, and Central and South America. They primarily hunt and eat large insects on the wing but will also prey on small birds, reptiles, and amphibians.

Fig. 2.23
Swallow-tailed Kite

Management skill level: Intermediate

Handling and husbandry: Swallow-tailed kites man well for program use and adapt nicely to a display, provided they have ample room. They appear most comfortable if housed with other Swallow-tailed kites or Mississippi kites. Their long tail feathers are prone to breaking if the birds are not managed properly.

Temperature: Swallow-tailed kites are warm weather birds that require an additional heat source if the temperature in their housing dips below 32°F (0°C).

2.2i Osprey

Osprey (*Pandion haliaetus*)

E

Description: Osprey are large, fishing-eating raptors that inhabit every continent except Antarctica. They have long wings, a dark body with light chest, a black eye stripe, and rough spicules on the bottoms of their feet for grasping slippery fish.

Management skill level: Expert

Handling and husbandry: Osprey are high-strung, easily stressed birds that do not tolerate proximity to people well. They are not recommended for program use and only those birds that eat well should be considered for display use. (They often do not eat well when first brought into captivity and some individuals never eat on their own while captive). In a display, they need to be provided with ample space and a large barrier from the public. Their primary flight feathers and tail feathers can easily break if they are not managed properly.

Medical: Osprey are susceptible to aspergillosis and should be put on a preventative treatment when first acquired, undergo a major management change, or become ill. They are also prone to developing bumblefoot, wrist injuries, and broken feathers due to their awkwardness in confined settings.

Temperature: Osprey are migratory in their northern range and must be provided with a heat source if the temperature in their housing dips below 32°F (0°C).

Fig. 2.24
Osprey

Fig. 2.25
Barred Owl

2.2j Owls

Barred Owl (*Strix varia*) ✩✩✩✩

Description: Barred owls are large tuftless, nocturnal owls that inhabit heavy mature forests often near lowland swamps in the eastern half of the United States, Washington, Oregon, and parts of Canada. They are brown mottled with white, have dark eyes, and a yellow beak. They prey on small rodents, birds, and crustaceans.

Management skill level: Intermediate

Handling and husbandry: Barred owls are relatively easy to man and handle a display setting well. Their dark eyes give them a soft expression, but certain individuals can be deceptively aggressive towards people when free-lofted. Also, barred owls are prone to breaking their delicate tail feathers if not managed properly and if they wear equipment (anklets and jesses), their feathered legs must be monitored for irritation. Barred owls bathe regularly.

Boreal Owl (*Aegolius funereus*) ✩✩✩✩

Description: Boreal owls are small, tuftless, forest-dwelling owls that most typically inhabit deciduous or coniferous forests in the very northern tip of the United States and into Canada. They are gray/brown in color, with a light face, yellow eyes, pale beak, and distinct dark spots on the tops of their heads. They prey primarily on small rodents (such as red-backed voles) and will sometimes eat small birds.

Management skill level: Intermediate

Handling and husbandry: Boreal owls are feisty owls that can be effectively manned and acclimated to a display. They are cavity nesters and if free-lofted should be provided with a shelter or nest box. If they wear equipment (anklets and jesses), their legs must be monitored for feather wear and skin irritation.

Fig. 2.26
Boreal Owl

Medical: Boreal owls have been known to contract avian malaria and should be housed in a mosquito-proof enclosure.

Temperature: Due to their northern natural range (and physical adaptations to handle cold temperatures), these owls do not handle warm climates well.

Burrowing Owl (*Athene cunicularia*) ✪✪✪✪

Fig. 2.27
Burrowing Owl

Description: Burrowing owls are small grassland owls that inhabit areas of dry vegetation with burrows, usually made by colonial rodents such as prairie dogs. They are found in the western part of the United States, Florida, Central and South America and prey primarily on small rodents and insects. They are brown mottled with white and are characterized by long legs and a round face with yellow eyes.

Management skill level: Intermediate

Handling and husbandry: Burrowing owls are shy birds that run into burrows to hide from danger. They can be easily manned but usually retain their shyness. Their lightly feathered legs are delicate and must be monitored closely for skin irritation if equipment (anklets and jesses) is applied. Burrowing owls can comfortably be put on display, as long as they are provided with a natural setting including a burrow or tunnel. If housed in pairs, they get territorial and should not be housed with other small birds as they may kill them.

Temperature: Burrowing owls are warm weather birds that migrate from their northern summer range. An additional heat source must be provided if the temperature in their housing dips below 32°F (0°C).

Fig. 2.28
Common Barn Owl

Common Barn Owl (*Tyto alba*) ✪✪✪

Description: Common barn owls are medium-sized owls that can be found on every continent in the world. They are characterized by a light body with tawny/buff markings and are freckled with dark specs. They have a distinct heart-shaped white face bordered in tan and dark eyes. Their vocalization is a raspy hiss instead of a hoot. They inhabit areas in the country with open fields and hedgerows for hunting primarily small rodents, and particularly like old buildings or large hollow trees for nesting.

Management skill level: Intermediate

Handling and husbandry: Common barn owls can be manned for program use and can be housed in a display. They appear most comfortable in a display if a hutch or shelter box is provided.

Temperature: Barn owls can easily overheat in warm temperatures and require supplemental heat if the temperature in their housing dips below 32°F (0°C).

Great Gray Owl (*Strix nebulosa*) ☆☆

Fig. 2.29
Great Gray Owl

Description: Great gray owls are large northern owls with a gray body, relatively small yellow eyes, a pale beak, a distinct facial disk, and a white "mustache." They inhabit the boreal forests of the upper United States and Canada and prey on small mammal species and some birds.

Management skill level: Advanced

Handling and husbandry: Great gray owls are typically nervous birds that do not handle manning or close exposure to large groups of people well. They will acclimate to a display, provided the enclosure is large enough to provide them with adequate space. They have long fragile tail feathers that easily break if inadequate housing is provided. If great gray owls wear equipment (anklets and jesses), their feathered legs must be monitored for irritation.

Medical: Great gray owls are highly susceptible to aspergillosis in captivity. They should be put on a preventative treatment when first acquired, undergo a major change in management, or become ill.

Great gray owls have small feet for the size of their bodies and are prone to developing bumblefoot if their diet, talon length, and perch options are not adequate. In addition, they are also highly susceptible to West Nile virus and should be housed in a mosquito-proof enclosure. Vaccination is recommended.

Temperature: Great gray owls do not handle warm climates well.

Long-eared Owl (*Asio Otus*) ☆☆

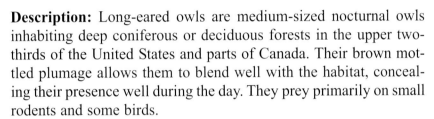

Description: Long-eared owls are medium-sized nocturnal owls inhabiting deep coniferous or deciduous forests in the upper two-thirds of the United States and parts of Canada. Their brown mottled plumage allows them to blend well with the habitat, concealing their presence well during the day. They prey primarily on small rodents and some birds.

Management skill level: Advanced

Handling and husbandry: Long-eared owls are high strung birds that are often nervous around large groups of people. With extreme patience and time, they can be manned for program use and do well in a display as long as habitat is provided to allow them to hide. If they wear equipment (anklets and jesses), their feathered legs must be monitored for irritation. Tethered birds have a tendency to break their tail feathers.

Short-eared Owl (*Asio flammeus*) ☆☆

Description: Short-eared owls are prairie owls, inhabiting open fields, meadows, and inland and coastal marshes. They are long-winged, buoyant birds that are buff/brown in color with barred wings and tail, yellow eyes circled in black, white/buff facial disks, and a black beak. They have dark wrist patches on their under wings, which helps identify them in flight. They are both diurnal and crepuscular hunters that feed on small mammals and sometimes birds.

Fig. 2.30
Long-eared Owl

Fig. 2.31
Short-eared Owl

Fig. 2.32
Snowy Owl

Management skill level: Advanced

Handling and husbandry: Short-eared owls are high-strung active birds that require time and patience for manning. They may or may not acclimate to large group exposure while on the glove, but generally will adapt to a display situation, as long as ample room is provided. If they wear equipment (anklets and jesses), their feathered legs must be monitored for irritation.

Snowy Owl (*Nyctea scandiaca*) ☆

Description: Snowy owls are large white northern diurnal owls inhabiting areas in the lowland tundra. Adult males are almost completely white with striking yellow eyes and a dark beak, while females have varying degrees of brown barring on their white plumage. They nest and perch on the ground. Snowy owls prey on a variety of animal species including arctic hares, lemmings, ground squirrels, grouse, ptarmigan, and shore birds.

Management skill level: Expert

Handling and husbandry: Snowy owls are high stress birds that are often flighty and nervous in captivity. They don't man well but may tolerate a display as long as a large space and low perches are provided. If equipment is applied (anklets and jesses), their feathered legs must be monitored for irritation.

Medical: Snowy owls are prone to bumblefoot in captivity. The condition of their feet must be closely monitored and management changes made as needed. In addition, snowy owls are extremely susceptible to aspergillosis in captivity. They should be put on a preventative treatment when first acquired, if any major management changes are made, and if they otherwise get ill or injured. Snowy owls are also highly susceptible to mosquito-borne diseases such avian malaria and West Nile virus. They should be housed with adequate mosquito protection and vaccination for West Nile virus is recommended.

Temperature: Snowy owls are very sensitive to warm temperatures and overheat easily. They should only be displayed in a northern climate and/or temperature controlled (cool, dry) environment.

2.2k Parabuteo

Harris's Hawk (*Parabuteo unicinctus*) ☆ ☆ ☆ ☆

N

Description: Harris's hawks are dark brown, medium/large hawks characterized by rust wing patches and leg feathers, and a dark chocolate tail with a thick white terminal band. They are found in semi-open arid areas of southwestern states and often hunt in family groups. Their diet consists of mammals, birds, and reptiles.

Management skill level: Novice

Handling and husbandry: Harris's hawks man well for program use. However, they are highly intelligent, social birds and can develop undesirable behaviors if poor, inconsistent training is provided. Once acclimated, they tolerate people and display situations well.

Medical: Harris's hawks can develop frostbite in cold temperatures and wing tip edema (swelling, fluid build-up) from any cause of circulatory compromise to the wings (including cold temperatures). They also are extremely sensitive to exhaust fumes and, like all other species, must always be housed in well-ventilated areas. In addition, they are highly susceptible to West Nile virus and should be housed in a mosquito-proof enclosure. Vaccination is recommended.

Temperature: Harris's hawks are warm weather birds and require an additional heat source if the temperature in their housing dips below 20°F (–6.7°C).

2.2l Vultures

Black Vulture (*Corapgys atratus*) ☆ ☆ ☆

A

Description: Black vultures are one of three North American vultures and may be more closely related to storks than raptors. They are social birds inhabiting open, slightly wooded, and urban areas in the southern United States, and Central and South America. Both

Fig. 2.33
Harris's Hawk

2.34
Black Vulture

juveniles and adults appear black, including their heads, and have pale whitish legs. In flight they can be distinguished from turkey vultures by the light undersides to their outer flight feathers. They are scavengers.

Management skill level: Advanced

Handling and husbandry: Black vultures can be manned for program use, but at times can be aggressive and bite on the glove. In general, they are more even-tempered than turkey vultures. They also can be housed in a display. If equipment is applied, their legs must be monitored for the build-up of caked feces or urates.

Temperature: Black vultures do not tolerate cold temperatures and must be provided with supplemental heat if the temperature in their housing dips below 32°F (0°C).

Turkey Vulture (*Cathartus aura*) ☆ ☆ ☆

Description: Turkey vultures are found throughout North, Central, and South America and Canada. Adults are most easily distinguished from black vultures by their red head but juvenile birds have a darker dusty head. They have a pink tinge to their feet and legs and all of their flight feathers are dark. The northern birds are migratory, not tolerating cold climates. They are scavengers.

Management skill level: Advanced

Handling and husbandry: Turkey vultures can be manned for program use, although some birds retain their defensive behavior of vomiting when approached by the handler or introduced to a group of people. They do well in a display but need close monitoring. They are "curious" birds that can exhibit destructive behaviors and ingest objects that are not part of their diet. Also, if equipment is applied (an anklets and jesses), their legs must be monitored for the build-up of caked feces or urates.

Medical: Some individuals lay down frequently and are prone to developing keel sores.

Temperature: Turkey vultures do not tolerate cold climates and

Fig. 2.35
Turkey Vulture

must be provided with a heat source if the temperature in their housing dips below 32°F (0°C).

2.3 SUMMARY

Before choosing a species and individual bird for your education program, you should first identify the goals of your program and think carefully about how a raptor will enhance the delivery of your messages. Then, you must consider the experience of your staff and the resources available to you to properly manage and care for a particular bird for years to come. It is critical to keep in mind that a bird's age, sex, and history all play a critical role in how it needs to be managed and the experience level required to successfully train and maintain it with a high quality of life. As a bird ages and reaches sexual maturity its behavior often changes and may require you to alter your training and management strategies (see chapter 8, Training).

TRC recommends American kestrels, Eastern screech owls, great horned owls, Harris's hawks, Northern saw-whet owls, red-tailed hawks, and Western screech owls for novice handlers. Accipiters and osprey are not recommended for program use.

If you are going to obtain captive-raised birds for education, contact the breeder for information on manning and management techniques of different species. Captive-bred birds often exhibit different behaviors than birds originally from the wild.

TRC recommends that all novice handlers/managers receive training from an experienced person or respected educational facility.

2.4 SUGGESTED READINGS

Burton, J. ed. 1992. *Owls of the World.* Oxfordshire, UK: Eurobook Ltd.

Ferguson-Lees, J. and D. Christie. 2001. *Raptors of the World.* Boston: Houghton Mifflin Company.

Johnsgard, P. 2001. *Hawks, Eagles, and Falcons of North America: Biology and Natural History.* Washington DC: Smithsonian Institution Press.

Johnsgard, P. 2002. *North American Owls: Biology and Natural History.* Washington DC: Smithsonian Institution Press.

Sparks, J. and T. Soper. 1996. *Owls.* Devon, UK: David and Charles.

Table 2.1 Symbol Legend

CATEGORY	SYMBOL	MEANING
RANKING		Overall rating of species based on training, handling, behavior, and management. One to five stars, with five stars being the highest rating — most recommended.
DIET		Mice
		Rats
		Day-old chicks
		Poultry (juvenile to adult ages)
		Rabbits
		Fish
		Quail
		Insects
PROGRAM AND DISPLAY USE		Recommended for program use
		Not recommended for program use
		Recommended for display use
		Not recommended for display use

HANDLING EXPERIENCE NEEDED	N	Novice (Have managed/trained 0-2 species.)
	I	Intermediate (Have managed/trained 2-3 species.)
	A	Advanced (Have managed/trained more than 3 species, including 1-2 intermediate level birds; minimum of 1 year bird management experience recommended.)
	E	Expert (Have managed/trained more than 6 species, including 1-2 advanced level birds; minimum of two years bird management experience, and recommendation from an expert handler.)
MEDICAL CONCERNS		The species has specific medical concerns in captivity. See chapter 7, Medical Care, for additional information.
TEMPERATURE ISSUES		The species has temperature sensitivities.

Table 2.2 Raptor species:

listed by level of recommended management experience

Novice	Intermediate	Advanced	Expert
American kestrel	Barred owl	Black vulture	Bald eagle
Eastern screech owl	Boreal owl	Long-eared owl	Cooper's hawk
Great horned owl	Broad-winged hawk	Merlin	Crested caracara
Harris's hawk	Burrowing owl	Northern harrier	Ferruginous hawk
Northern saw-whet owl	Common barn owl	Peregrine Falcon	Golden eagle
Red-tailed hawk	Red-shouldered hawk	Prairie Falcon	Great gray owl
Western screech owl	Rough-legged hawk	Short-eared owl	Northern goshawk
	Mississippi kite	Swainson's hawk	Osprey
	Swallow-tailed kite	Turkey vulture	Sharp-shinned hawk
			Snowy owl

Chapter 3: DIET

Wild raptors feed on a wide variety of vertebrate and invertebrate prey. Everything from fish to snails, rabbits to mice, and a wide variety of insects, birds, amphibians, and reptiles is eaten by some raptor at some time. Species, as well as individuals, vary in the type and variety of food they consume. Some species are generalists, eating a wide range of prey, varying their diet based on what is available. Others are specialists, limiting themselves to one or a few kinds of animals. Most raptors also take advantage of carrion if they come upon it in the wild.

When maintaining raptors in captivity, it is desirable to imitate their natural diet as closely as possible. This includes providing the type, as well as the variety, of prey they might naturally consume. TRC believes strongly that your bird's physical and mental health will benefit from this approach.

3.1 WHAT TO FEED

A diet of raw meat such as hamburger, chicken breast, or organ meat (heart and kidney) is not a proper diet for any raptor. This is a very important point, so it will be repeated:

It is never enough to feed just raw meat to a raptor!

Also, you cannot feed a raptor a primarily raw meat diet and supplement it sufficiently to make it wholesome. Raw meat is nutritionally imbalanced, with an excess of phosphorus that dangerously affects the amount of usable calcium (chapter 7, Medical Care). In addition, some people have tried to maintain raptors on a diet primarily of chicken necks because the necks were inexpensive and easy to acquire. A chicken neck diet will result in a life-threatening vitamin B deficiency.

Captive raptors must be fed whole animals, usually birds or mammals, or a mixture of whole prey animals and commercial diets. TRC prefers using the more natural whole-animal diet. When feeding a natural diet, it's important to include the flesh, organ meat, viscera (of freshly killed prey items), bones, and skin of the prey animal. This ensures that your raptor is getting the proper balance of vitamins, minerals, and casting materials as well as the calories it needs to survive.

This is probably as good a time as any to bring up the subject of pellet egestion, or casting. Certain parts of any raptor's natural diet are indigestible. This includes such things as fur, feathers, scales and, in the case of owls, bones. To get rid of these materials, raptors cast pellets. Pellets are compact bundles of indigestible material formed in the stomach of the birds and regurgitated, usually daily. You should look for these pellets, as they are a good indicator that your bird is eating and processing food normally.

3.1a Domestic Birds and Mammals

To achieve your goal of a balanced diet, feed a variety of freshly killed rodents and/or poultry. Mice, rats, rabbits, hamsters, guinea pigs, chicken, turkey (of all ages), and quail are all commonly used as food for captive raptors. If feeding the larger mammal and avian items (rabbit, adult chicken, etc.) do not offer your bird the larger bones. In their zealousness for dinner, many birds will eat rapidly, ingesting large pieces that can easily get lodged in the crop, or esophagus. Also, be cautious of feeding poultry necks as these also can be ingested whole, fold over in the crop or esophagus, and get stuck.

3.1b Pigeon

Another prey item people feed to raptors is pigeon. TRC strongly cautions you, however, against doing this. Pigeons are known carriers of a variety of fatal diseases, such as frounce (trichomonas), herpes virus, Newcastle disease virus, avian tuberculosis, and psitticosis (chapter 7, Medical Care). Frounce cannot be prevented by removing the head, neck and digestive tract of a pigeon (the liver and pectoral muscles are often contaminated) but it can be avoided by completely freezing and then thawing a pigeon prior to feeding. However, the freezing process does not eliminate the other diseases listed. To be safe, TRC does not feed pigeons, either wild or domestically raised, to any of our raptors.

3.1c Wild Food Sources

You must be careful if feeding your raptor "wild" food sources. Any wild species should be in good physical condition, killed by a known acceptable method, and be properly and quickly preserved or fed fresh. Animals that should definitely be avoided include those that are shot with lead (they can easily cause fatal lead poisoning in your bird), rodents that are just found dead (these were probably sick or potentially poisoned), and road-kill that is not fresh (in the summertime, unless you know a road kill is fresh, drive on past). There is always the risk that a wild animal harbors lead, parasites, carries diseases, or in the case of road-kills, have a flourishing clostridium (bacterial) load that could be harmful to your bird. So, be cautious.

Another issue revolves around feeding English sparrows (*Passer domesticus*) and European starlings (*Sturnus vulgaris*). Both are non-native, unprotected species. These species are carriers of West Nile virus and there is always the risk that your bird could become infected by eating infected individuals. The possibility is greatest from mid summer to mid fall, when the English sparrows and European starlings are harboring the greatest number of viral particles in their systems. These birds carry the virus without getting sick, so you cannot determine the potential threat by looking at them. Also, the freezing process does not necessarily kill/disable the virus. Therefore, TRC cautions you against feeding them. There is a lot that is not yet known about this virus and there is always a risk. For the most up-to-date information, contact The Raptor Center.

3.1d Immature Animals

Although research has shown that some immature animals, such as day-old poults, are nutritionally complete, TRC does not recommend feeding your raptor a diet exclusively of immature animals. Raptor managers often like the young animals because they can be acquired cheaply and easily. However, to ensure that your bird is getting all the nutrients it needs, and to vary the type, texture, and size of prey items a little to enrich your bird's life (chapter 4, Housing), TRC recommends that 50 percent or less of your bird's diet be made up of poultry chicks or baby mammals.

3.1e Commercial Diets

Commercially prepared diets are available (appendix B). TRC does not recommend using these as the sole food source. A variety of whole animals should still be included in any captive raptor's diet. An improper diet can cause nutritional disorders (chapter 7, Medical Care).

The bottom line is that it is critical to your bird's health to mimic its natural diet as much as possible. It is not appropriate to feed a bird-eating raptor strictly rodents (or a rodent-eating raptor strictly birds). A bird-eater should receive a diet that consists of at least 50 percent bird prey items. It is not able to extract as much energy and nutrition from rodents, which often leads to problems with its overall health, beak and talon condition, feather quality, and health of its skin. Do your homework ahead of time to learn what your bird would eat in the wild and how you are going to provide it with the best diet possible.

3.2 HOW MUCH TO FEED

As a general rule of thumb, 100–200 g (4–8 oz) raptors will eat 20–25 percent of their body weight daily, 200–800 g (8 oz–1.8 lb) raptors will eat 15 percent of their body weight, 800–1,200 g (1.8–2.8 lb) birds will eat 10 percent of their body weight, and raptors larger than 1,200 g (2.8 lb) will eat approximately 6–8 percent of their body weight daily. This amount varies with age, activity level, weather, and even time of year. By monitoring your bird's weight (chapter 6, Maintenance Care) and behavior, you can adjust the quantity of food offered. Table 3.1 lists the recommended guidelines for feeding program and display raptors.

Birds can lose weight, become anemic, and die if not fed a sufficient amount of highly nutritious food; they can also become obese if provided with an overabundance of food. Obesity is a medical condition that results in potentially fatal health problems, such as fatty liver disease (chapter 7, Medical Care). Therefore, it is critical to weigh your bird regularly, monitor its annual/seasonal consumption patterns, and adjust the amount of food offered to keep your bird in a healthy weight range.

3.3 WHERE TO OBTAIN FOOD

After looking at table 3.1, you can see that the recommended staple diet consists of a mixture of rats, mice, poultry, and coturnix quail. The rodents can either be purchased from a variety of commercial breeders, or sometimes acquired at no charge from laboratories. If getting rodents from a lab, be sure to limit yourself to "control" or surplus animals, to avoid drugs or other chemicals that might get passed on to your bird. Also, inquire into the method of euthanasia used. Potential food animals should not be injected with euthanasia solutions (e.g. barbiturates); they can cause secondary poisoning in your raptor. In addition, check the rodents carefully for any markers, such as metal ear clips, that need to be removed before feeding them to your bird.

Poultry and quail can also be purchased from commercial breeders and sometimes hatcheries have day-old chicks available at no charge. A list of some food suppliers is presented in appendix B. Before acquiring food from any source, do a little research to make sure the animals are raised and euthanized humanely, and are frozen quickly.

You can also raise your own food; depending on the quantity you need and the facilities you have available, this can be an attractive alternative. One important rule to follow is never feed any animals to your raptors if they exhibited signs of illness. With age, rodents sometimes get sick and develop tumors or excessive weight loss. These animals should be euthanized and discarded.

If you raise or receive live food, it should be euthanized humanely. Killing animals is never a pleasant task; however, if you keep a raptor, it has to be done, either by you or by someone else. As noted above, you can't use any chemical that might be passed on to your bird. The University of Minnesota's approved method of euthanasia for small rodents is a CO_2 chamber.

3.4 HOW TO STORE FOOD

Carcasses can be frozen and stored for up to six months with little loss of quality (except for fish, which lose vitamins quickly). Most people who keep raptors have a freezer just for storing the birds' food. It is critical that when freezing food, the food is laid flat in a thin layer to avoid rotting and bacterial contamination.

Food can be packaged in meal-size servings and thawed when needed. Thawing whole animal carcasses in cool/cold water is best to avoid bacterial contamination and to replace a little moisture that was lost in the freezing process.

3.5 HOW TO FEED

3.5a Preparation

Just as the proper presentation of a meal at a four-star restaurant is important to your gourmet-dining experience, so too is the presentation of food important to the discerning raptor. Some raptors, like some people, will eat almost anything, any time, anywhere, but most will eat better if their food is prepared and presented with a little thought.

Contrary to popular belief, captive raptors don't need to be fed live prey. In fact, attempting to do so could cause injury to your hawk or owl depending on its housing facility and physical limitations. Except in extraordinary situations, food should be presented to your raptor only after it has been euthanized.

Food should always be thawed and presented in a fresh-looking manner. Food that is frozen or even too cold can cause digestive problems such as "sour crop," which can lead to serious sickness (chapter 7, Medical Care). The intestines of previously frozen rodents should be removed, as they tend to rot quickly, the birds won't eat them, and they can be a potential carrier of clostridium (a deadly bacterium). The liver, however, is almost always a welcome, very nutritious delicacy and should be offered. Lying open the belly of the prey to get to the intestines will also expose the red organ meat, an irresistible sight to even the most jaded raptor.

Preparing the food in this fashion should be done immediately prior to offering the food. Freshly killed animals can be fed whole (including the intestines) as can day-old chicks if fresh or quickly frozen and thawed in cold water. The yolk and digestive system are very nutritious and the birds seem to like them. Food that has suffered freezer burn, or looks or smells "bad," should never be fed to your bird.

3.5b Vitamins

If a bird is housed outdoors and is on a varied diet of fresh, whole prey items, vitamins should not be necessary. However, most raptor caretakers feed their birds previously frozen food (which has lost moisture and vitamins during the freezing/thawing process) and provide their birds with a relatively short menu. To make sure your bird is receiving the highest quality nutrition possible, a vitamin supplement can be added to the diet.

When choosing a supplement, keep in mind that vitamin compounds are made with a specific type of animal in mind (human, canine, feline, etc.). If you give your bird an incorrectly balanced supplement (one not made for hawks) you run the risk of deficiencies or excess absorption of one vitamin, both potentially harmful situations. Vitahawk Maintenance®, a multivitamin powder made specifically for hawks, has been used safely for several years and is available through a variety of sources. If you are thinking about using a different vitamin, speak with your veterinarian to see if it will be appropriate for your bird. Make sure to follow the dosing guidelines provided on the packaging. Vitamin overdoses can lead to serious health problems (chapter 7, Medical Care).

One vitamin that is not available to a raptor through its diet is vitamin D3. Birds normally acquire this essential vitamin when ultraviolet irradiation (from sunlight) of the oil in their feathers converts vitamin D2 (which cannot be absorbed) into vitamin D3. When birds preen their feathers, they ingest the vitamin D3. Birds housed outdoors should have a sufficient amount of vitamin D3 to fill their dietary requirements. However, if a bird is housed indoors for a lengthy period of time and has little to no access to sunlight (TRC does not recommend this), a vitamin D supplement may be necessary.

Vitamin D is essential for the absorption of calcium and phosphorus from the digestive tract and for regulating the amount of calcium circulating in the blood. One vitamin D supplement often given is cod liver oil. A drop or two of cod liver oil added to your bird's food once per week is all you need. However, vitamin D is a fat-soluble vitamin and if overdosed can cause severe health problems. Therefore, TRC does not recommend using it if multiple birds are housed together (one bird may get more than its share) and strongly encourages people to provide their bird with regular direct sunlight instead.

Fish that has been frozen for more than a few months loses essential vitamins, particularly thiamine. For raptors that eat a strictly fish diet, such as osprey, adding a thiamine supplement (10 mg/kg once per week) is important to prevent nervous system disorders.

One last note on vitamins: supplements are not a substitute for a high quality diet.

3.5c Delivery

Whenever feeding your bird, there are a few safety tips to follow. Keep in mind that most raptors look forward to dinnertime and sometimes get aggressive for their food.

- Birds that are fed on the glove should be out of site of other birds during feeding time and fed before other birds. Otherwise, they get anxious and aggressive for their food.

- If feeding a bird on the glove, the food should be presented at foot level. Do not reach up and present the food to the bird's beak. Forceps can be used to bring food to the glove to minimize the risk of footing. Routinely offering your bird food from your hand teaches your bird that your hand means food and it will reach out to grab your hand with its feet or beak. The ONLY time birds may see food in your bare hand is during specific training scenarios (chapter 8, Training).

- If birds are just left food in their enclosure, it is a good idea to wear a glove on the hand carrying the food container and/or handling the food. Raptors learn when they are going to be fed and over-eager eaters can respond quickly to the presence of their caretakers at that time. They may fly at the person leaving food, and/or reach out to grab the food, accidentally grabbing a bare hand.

- Food should be hidden from view until it is placed in the feeding location.

It's important to clean up uneaten food. Spoiled food can lead to salmonellosis or other bacterial diseases. Many species of raptors "cache," or hide, food. This natural behavior serves them well in the wild by letting them save uneaten and excess prey for a day or two. It's unnecessary in captivity, however, and attracts flies, ants, bees, rodents, and bacteria as well as other predators.

3.6 WHERE TO FEED

If you are feeding your bird in its enclosure, placing the food in the same spot every day will encourage eating. Raptors tend to be creatures of habit, and will more easily recognize the offered food as dinner (especially if it's different or unusual) if they see it in the same place every day. If two or more birds are housed together, make sure to place their meals in separated piles to prevent aggression. Also make sure neighboring raptors are either fed at the same time or have a visual barrier so they don't see the food.

Food should be placed on a feeding platform or perch. Do not place food directly on gravel or sand. These substrates stick to the food and can easily get ingested, potentially causing obstructions.

3.7 WHEN TO FEED

Raptors rarely need to be fed more than once a day. In extremely cold weather, however, small birds such as American kestrels, Northern saw-whet owls, and Eastern and Western screech owls may require food twice a day if they are housed outdoors. In mild weather, larger birds such as great horned owls, red-tailed hawks, bald eagles, and golden eagles could be fed once every other day, but TRC believes in feeding them everyday as a method of enrichment.

Also, during warm weather, one fasting day per week can be incorporated into the feeding regime of the larger species to help control their weight (they are often eager eaters and can easily get obese if their weight is not monitored carefully). Tempting as it may be, this does not mean you should feed them a little extra the day before or after. TRC does not recommend any fasting days for imprinted birds, or those with destructive tendencies such as caracaras and vultures. These birds are easily bored and need as much stimulation as possible.

Wild raptors generally feed in the morning or late afternoon. While it's certainly not essential to feed at those particular times, establishing a routine will keep your bird on a regular casting cycle as well as encourage eating. During daylight saving time, TRC feeds all its education birds in the late afternoon, with the owls fed as late as possible to mimic their normal behavior patterns in the wild. When the days get short, the hawks are fed in the early afternoon and the owls in the later afternoon, close to dark.

3.8 WATER

Your bird should always have a source of water available, except during freezing temperatures. Many birds drink frequently, especially falcons, eagles, and barred owls. In cold climates, it is a good idea to periodically bring your avian educator indoors, if possible, and give it the opportunity to drink and bathe. Make sure the bird and its equipment are completely dry before it returns to its frigid outdoor enclosure. Keep in mind that when housing a raptor in captivity, you control the amount of food it receives, and you also offer it limited resources for adapting to weather conditions and stressful situations. Therefore, your captive bird will probably need more water than a bird in the wild, and you might find that it readily drinks the water provided. See chapter 4, Housing for more details.

3.9 THE RELUCTANT EATER

Normally, raptors eat very willingly in captivity, so getting your bird to eat shouldn't be difficult. For various reasons, however, captive raptors sometimes refuse to eat. It's not unusual for birds that have been recently moved (newly acquired birds or those moved to a different mew), or had some other change in their management to temporarily stop eating. However, usually within a few days, their hunger overrides their nervousness and temptation wins. Eastern screech owls are notorious for not eating after a change in management. Listed below are a few suggestions TRC has found useful for getting a reluctant bird to eat.

3.9a A Newly Acquired Bird

- Make sure the area is quiet, with as little disturbance as possible. Constant distractions may prevent a bird from concentrating on its food.

- Try a variety of food items to see if one is preferred or recognized. The commercial bird-of-prey diet is seldom accepted right away, because raptors do not recognize it as food. If you want to include this food in your bird's diet, you might need to mix it in with the recognizable food items.

- If different species are housed next to each other, make sure they have a visual barrier. Sometimes, a bird won't eat if it feels intimidated by the sight of another raptor, especially a potential predator.

• If possible, temporarily move the bird next to or in with another bird of the same species or of a compatible species (table 4.5) that is a consistent eater. Often the sight of another bird eating will stimulate a non-eater's appetite.

• If the bird is housed with other birds, the other residents may intimidate it, it may not be aggressive enough for the food, and/or the other birds may be possessive of the food and prevent the new tenant from eating. Isolate the newcomer and once it eats, start a slow introduction back into the enclosure (chapter 8, Training).

• If the bird will allow you to approach it, try offering juicy pieces of meat such as liver and heart; this might stimulate your bird's appetite in its new environment.

3.9b A Resident Undergoing Management Change

(new location, new handler, etc.)

• Make sure the new area is quiet, with as little disturbance as possible.

• Make sure there are visual barriers to other raptors or animals. Other animals may not only be distracting, but threatening.

• With a location change, make sure all other management practices remain unchanged. This would include the time of feeding, the person who feeds, the type of food offered, and the location of food placement (feeding platform, perch, etc).

• If necessary, conduct a slow transition to a new location. Place your bird in the new enclosure for part of a day but then return it to its old enclosure, if possible, to feed.

• If the bird is routinely fed off the fist but won't eat with a new handler, after a day or two give it a break and either leave the food in the bird's enclosure or have the old handler feed it while the new handler is close. A slow transition may be required.

3.9c A Resident Not Undergoing Management Change

If your bird has not undergone any change in location or management and stops eating for two days, consult your veterinarian. Sometimes, if the weather is hot a bird may go off food for a few days. However, once a bird is acclimated to its surroundings and captive life, it usually eats at least partial meals daily, even in warm weather. If it doesn't, it could mean your bird is sick and you should not waste any time diagnosing the problem. Remember that a bird that doesn't eat for a few days (and doesn't drink water) can become dehydrated, which only compounds the anorexia and potential illness.

Birds that lay eggs also often stop eating and/or have a greatly reduced appetite. This situation must be monitored closely. Contact your veterinarian and/or falconers or zoos with captive breeding experience for information and guidance.

3.10 RECORD KEEPING

It is important to keep good records of your bird's daily eating patterns. Annual feeding patterns and weight ranges do exist and once you establish what is normal for your bird, you can relax a little when your bird shows expected changes in appetite. For example, captive eagles, and some of the larger hawk and owl species, may lose as much as 200g in the spring (in the Midwest), and gain this weight back to bulk up for the winter. Figure 6.1 shows the forms used by TRC to record and monitor the weights of education raptors.

3.11 SUMMARY

A diet consisting exclusively of raw meat is never sufficient for any raptor. To ensure proper nutrition, your bird's diet must also contain skin, bones, and organ meat.

Finding a reliable supply of high-quality food in sufficient quantities is one of the most important tasks in caring for a captive raptor. It can be expensive and time-consuming and is a never-ending proposition. If the money is available, you can purchase food from a number of sources. Food can be kept frozen for up to six months, but must be properly and completely thawed before it is fed to your bird. While some birds eat regularly and with no prompting, others, particularly birds moved to new facilities, might need special treatment. In general, a quiet and peaceful setting, along with established routines, will ensure that your bird will eat heartily and well. It is essential to monitor and record your bird's eating patterns throughout the year so you can be prepared for shifts in appetite and food consumption and identify potential problems early.

3.12 SUGGESTED READINGS

Bent, Arthur. 1972. *Life Histories of North American Birds of Prey. Part One.* New York: Dover Publications, Inc.

Bent, Arthur. 1972. *Life Histories of North American Birds of Prey. Part Two.* New York: Dover Publications, Inc.

Craighead, J. and F. Craighead. 1977. *Hawks, Owls, and Wildlife.* New York: Dover Publications Inc.

Ferguson-Lees, J. and D. Christie. 2001. *Raptors of the World.* Boston: Houghton Mifflin Company.

Forbes, N. and C. Flint. 2000. *Raptor Nutrition.* Worcestershir, UK: Honeybrook Farm Animal Foods.

Johnsgard, P. 2002. *North American Owls, Biology and Natural History.* 2nd ed. Washington DC: Smithsonian Institution.

Johnsgard, P. 2001. *Hawks, Eagles, and Falcons of North America.* Washington DC: Smithsonian Institution.

Table 3.1 Recommended average daily food rations for program and display raptors *

Species	Sex	Amount	Food Type **
American kestrel	Male	20-25g (0.7-0.9oz)	**Mice**, chicks, quail, insects
	Female	25-30g (0.9-1.1oz)	
Bald eagle	Male	200g (7.1-10.6oz)	**Fish, rat**, poultry, quail, rabbit
	Female	300g (10.6oz)	
Barred owl	Male	50g (1.8oz)	**Mice**, chicks, quail, rat
	Female	80g (2.8oz)	
Black vulture	Male or Female	100g-125g (3.5-4.4oz)	**Rat**, chicks, fish, poultry, quail, rabbit
Boreal owl	Male	30g (1.1oz)	**Mice**, chicks, quail
	Female	35-50g (1.2-1.8oz)	
Broad-winged hawk	Male	30-40g (1.1-1.4oz))	**Mice**, chicks, quail, rat
	Female	40-50g (1.4-1.8oz)	
Burrowing owl	Male or female	50g (1.8oz)	**Mice**, chicks, insects
Common barn owl	Male	50g (1.8oz)	**Mice**, chicks, rat
	Female	80g (2.8oz)	
Cooper's hawk	Male	40-50g (1.4-1.8oz)	**Quail**, chicks, mice, rat
	Female	50-60g (1.8-2.1oz)	
Crested caracara	Male or Female	100-140g (3.5-4.9oz)	**Mice**, chicks, poultry, quail, rat
Eastern screech owl	Male or female	20-30g (0.7-1.1oz)	**Mice**, chicks, insects, quail
Ferruginous hawk	Male	100-125g (3.5-4.4oz)	**Mice, rat**, chicks, poultry, quail, rabbit
	Female	125-150g (4.4-5.3oz)	
Golden eagle	Male	200g (7.1oz)	**Rat, poultry**, quail, rabbit
	Female	300g (10.6oz)	
Great gray owl	Male	60-80g (2.1-2.8oz)	**Mice**, chicks, quail, rat
	Female	80-100g (2.8-3.5oz)	
Great horned owl	Male	60-80g (2.1-2.8oz)	**Rat, mice**, chicks, poultry, quail, rabbit
	Female	80-120g (2.8-4.2oz)	
Harris's hawk	Male	60-80g (2.1-2.8oz)	**Mice, chicks**, quail, poultry, rabbit, rat
	Female	80-100g (2.8-3.5oz)	

Long-eared owl	Male or female	50g (1.8oz)	**Mice**, chicks, quail
Merlin	Male	25-30g (0.9-1.1oz)	**Quail**, chicks, mice
	Female	30-40g (1.1-1.4oz)	
Mississippi kite	Male or female	35-40g (1.2-1.4oz)	**Mice**, chicks, insects, quail
Northern goshawk	Male	80g (2.8oz)	**Quail**, chicks, poultry, rab, rat
	Female	100g (3.5oz)	
Northern harrier	Male or female	50-60g (1.8-2.1oz)	**Mice**, chicks, quail
Northern saw-whet owl	Male	15-20g (0.5-0.7oz)	**Mice**, chicks, insects, quail
	Female	20-25g (0.7-0.9oz)	
Osprey	Male	150-175g (5.3-6.2oz)	**Fish**
	Female	175-200g (6.2-7.1oz)	
Peregrine falcon	Male	80-100g (2.8-3.5oz)	**Quail**, chicks, mice, poultry, rat
	Female	100-120g (3.5-4.2oz)	
Prairie falcon	Male	80-100g (2.8-3.5oz)	**Quail, rat**, chicks, mice, poultry
	Female	100-120g (3.5-4.2oz)	
Red-shouldered hawk	Male	50g (1.8oz)	**Mice,** chicks, quail, rat
	Female	60-80g (2.1-2.8oz)	
Red-tailed hawk	Male	60-80g (2.1-2.8oz)	**Mice, rat**, chicks, poultry, quail, rabbit
	Female	80-100g (2.8-3.5oz)	
Rough-legged hawk	Male	60-80g (2.1-2.8oz)	**Mice**, chicks, quail, rat
	Female	80-90g (2.8-3.2oz)	
Sharp-shinned hawk	Male	25-30g (0.9-1.1oz)	**Quail**, chicks, mice
	Female	30-40g (1.1-1.4oz)	
Short-eared owl	Male or female	50g (1.8oz)	**Mice**, chicks, quail
Snowy owl	Male	125-150g (4.4-5.3oz)	**Mice, rat, poultry,** chicks,quail, rab, rat
	Female	150-200g (5.3-7.1oz)	
Swainson's hawk	Male	60-80g (2.1-2.8oz)	**Mice**, chicks, insects, quail, rat
	Female	80-100g (2.8-3.5oz)	
Swallow-tailed kite	Male or Female	40-50g (1.4-1.8oz)	**Mice**, chicks, insects, quail
Turkey vulture	Male or Female	100-125g (3.5oz-4.4oz)	**Rat**, chicks, fish, poultry, quail, rabbit
Western screech owl	Male or female	20-30g (0.7-1.1oz)	**Mice**, chicks, insects, quail

*Please note that these quantities are designated for raptors that are in good weight, not undergoing weight management for free flight, and are housed in temperate climates. Also, keep in mind that the size of species varies with geographical regions. With the exceptions of the black vulture, crested caracara, kites, and Western screech owl, the quantities listed here are for members of the species living in the upper Midwest.

**The first item(s) listed in bold are typically the staple diet of each species. The other items, listed alphabetically, are a selection of additional food types each species will typically eat. Offering items from these lists twice per week will help provide your bird with a healthy diet.

Chapter 4: HOUSING

There are many important factors involved in keeping a raptor in captivity. Along with food, housing is one of the most basic and, to the novice, most intimidating of these factors. As with human houses, the varieties and styles of raptor housing can seem endless. The task of providing suitable housing will be much easier if you keep in mind that it must do three basic things: prevent your bird from escaping, protect it from predators, and give it shelter from the elements. The real difficulty lies in doing these three things while keeping your bird physically and mentally healthy.

The guidelines presented in this chapter apply to raptors that are used for education program or display use. Raptors held for the sport of falconry, captive breeding programs, or free-flight demonstrations have different housing requirements that are not addressed here.

Due to the specialized needs of captive raptors, there are no ready-made raptor houses. Cages made for parrots, dogs, or other animals will not work. Therefore, you will need to build a suitable structure from scratch or modify an existing raptor enclosure for a new bird with special needs. For purposes throughout this book, an enclosed facility is referred to as a mew and an outdoor area as a weathering yard.

It's important to give the housing you create — including the design, materials, construction, and maintenance — a great deal of thought, or it can end up being the source of endless problems. What are some of the important considerations that affect the type of housing you end up with? Here's a partial list of things to consider:

- Will your mew be indoors or outdoors?
- Is water/drainage close by?
- What is the size of your bird?
- Will your bird be on public display?
- Will your bird be handled for programs?
- Will your bird be free-lofted or tethered?
- How will the housing provide enrichment for your bird?
- To what temperature extremes will your bird be exposed?

- What is the prevailing wind direction?
- How safe is the area from predators?
- How safe is the area from human disturbance?
- Will your mew be easy to clean?
- What features will you install to present accidental escape?

4.1 LOCATION

Although raptors can be housed indoors or outdoors, keeping your bird outdoors is preferable. This exposes the bird to fresh air, sunshine, and rain, all of which are critical for your bird's long-term health. It also allows the natural elements to help keep pathogens at bay and helps to keep the smell of the bird's mutes and food away from people. An outdoor facility can also be versatile in different types of weather if built with features of both a mew and weathering area.

With these things in mind, your first decision will be where to put the mew. If the mew is outside, its location should provide protection from elemental extremes, such as wind and hot sun, while at the same time taking advantage of cooling breezes in hot weather and warming sun in cold weather. A mew is often constructed to face east to meet these requirements. It should also be placed away from other species of animals, especially those bedded with straw or hay; these substrates get moldy when wet and can lead to an increased number of aspergillus spores in the air (chapter 7, Medical Care). Also, consideration should be given to provide visual separation both from other species of birds and different kinds of animals.

Human traffic patterns are also an important consideration. You'll want to keep the mew a respectable distance from heavy machinery, recreational activities, and large groups of people. At the same time, the mew should be easily accessible to the bird's handlers and feeders. We recommend locating the mew in a fenced area that is away from the main entrance or front door of your building, doesn't allow people to walk around, above, or below it, isn't located next to noisy activities or animals, and is not close to exhaust fumes.

4.2 ENCLOSURE SECURITY

TRC strongly recommends taking precautions to protect your outdoor facilities from potential human intruders. Constructing the mew in an area where security lighting is available (motion detectors work well — light is not on all the time, just when triggered), and the public does not have easy access during non-visiting hours (in a fenced in area, courtyard, etc.) will help. It is also strongly recommended to lock all raptor enclosures as an extra precaution.

4.3 STRUCTURE

Creating appropriate housing for your bird isn't an insurmountable task. Building an enclosure that is both a mew and weathering area can simplify the process. Such a structure can be modified for different climates and birds, and can work for both free-lofted and tethered birds. Features of a combined enclosure will be described below. In appendix C, you will find photographs of enclosures that have worked well for raptor caretakers across the country.

4.3a Size and Shape

The enclosure can be a simple square or rectangle or it can be modified to take on the shape of the area it occupies. The appropriate cage size will vary with the size of your raptor (table 4.1), the number of birds occupying it, and any special needs your bird has (for example, partial wing amputees often need a ramp system to have access to a perch off the ground). At a minimum, the cage must be large enough to allow the bird to fully extend its wings without damaging its feathers or wrists. If a bird is tethered, it must have enough room to reach the end of its leash without hitting its wings on the walls. It is TRC's belief that a raptor should be housed in an area that is at least large enough to allow it to move between two perches (or to the top of a hutch if tethered). If a bird is forced to perch consistently in one spot, it will more than likely develop bumblefoot (chapter 7, Medical Care).

Use common sense when designing your raptor cage. Don't plan a fifty-foot long (15.2m) enclosure for an American kestrel or a three-foot long (0.9m) enclosure for a red-tailed hawk. The kestrel will be difficult to retrieve for programs or medical exams and the red-tailed hawk will not have enough room to move around, be comfortable, and feel secure. Both situations would result in additional stress for your bird. Keep in mind that as a raptor caretaker, it is your responsibility to provide the highest quality of life possible for the bird(s) in your care.

4.3b Walls

The back wall should be solid, to protect the bird from the elements and give it a feeling of security. Untreated plywood works well. Avoid pressure treated plywood as it contains chromated copper arsenate, which can be harmful to your bird if ingested. In

Fig. 4.1
Types of vestibules. (a-c Gail Buhl)

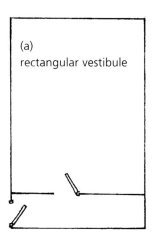

(a)
rectangular vestibule

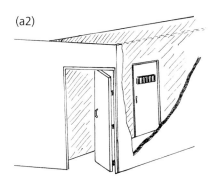

(a2)

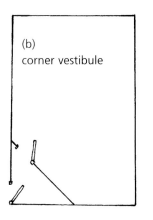

(b)
corner vestibule

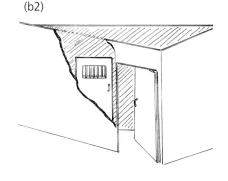

(b2)

addition, TRC has found that having the sidewalls solid for at least the back third or half of their length provides additional protection from the elements and gives the bird a greater feeling of security. If for some reason this is not practical, make at least one back corner solid. The remainder of the sidewalls, along with the front wall, can be constructed of nonsolid materials: welded wire, heavy-duty plastic mesh, vinyl coated chain link, slats made of wood or Trex® decking material (this product is made of recycled plastics and is resistant to mold, weathering, and termites), or dowels (wood or conduit).

If slats or dowels are used, TRC recommends putting welded wire around the outside to prevent accidental escape or unwanted visitors. The space between dowels or slats should be small enough to prevent a bird from escaping or from getting a foot caught between them. Dowels should be spaced such that the area between them is no greater than ½ inch (1.3 cm) apart for small and medium-sized birds (American kestrels, Northern saw-whet owls, long-eared owls, red-shouldered hawks, etc), 1 inch (2.5 cm) apart for large birds (barred owls, great horned owls, red-tailed hawks, etc), and 2 inches (5.1 cm) apart for extra large birds (eagles).

Don't use chicken wire in raptor caging. It will damage your bird's feathers, cere, and feet if your bird flies into the wire. If you use a mesh, make sure the holes are small enough to prevent escape of free-lofted raptors and to prevent your bird from being able to put its foot through the mesh holes. Mesh with ½-inch (1.3 cm) holes for small and medium-sized birds, 1¼-inch (3.2 cm) holes for large birds, and 2-inch (5.1 cm) holes for extra large birds (eagles) works well. This allows for viewing the bird as well as exposure to sun and fresh air. You can buy extruded heavy-duty mesh (high density polyethylene) from some home-improvement stores, bird-control product suppliers, and net companies.

Due to the threat of West Nile virus and, in some areas, avian malaria, TRC advises covering the outside of your raptor mew with mosquito netting. West Nile virus is a concern for all species and you should check with your local health department to find out the prevalence of the virus in your area. Avian malaria, a deadly disease caused by a blood parasite, is also transmitted by mosquitoes. It has been known to affect a variety of species including gyrfalcons, merlins, eastern screech owls, and American kestrels. Additional information on these diseases can be found in chapter 7, Medical Care.

4.3c Doors

The door of the enclosure should be along the front or side wall, and should be either solid (plywood) or made of welded wire, vinyl coated chain link, heavy duty plastic mesh, dowels, or slats. TRC highly recommends a vestibule, or two-door system, to reduce the possibility that your bird will accidentally escape while you are coming and going (figure 4.1a, b).

If such a vestibule isn't possible, there are several design modifications you can make to prevent accidental escape. One is to construct your door half the height of the cage. You'll need to bend when entering, but it will help discourage a free-lofted bird from escaping when the door is opened (most raptors don't like to fly low). However, this system is not recommended for species that like to reside on the ground such as vultures and caracaras.

Another modification is to place perches higher than the door; this will provide your raptor with a secure spot to perch, and it will be less likely to swoop down and out.

Another tip, if a vestibule is not possible: place the door near a corner of the cage and have it open into the cage (figure 4.1c) instead of away. This will make it more difficult for a bird to zip past you.

Finally, a portable vestibule can be constructed out of PVC pipe and game bird netting (figure 4.1d). If it is not possible to secure the unit against the outside wall it should only be used for small birds that would be unable to move the unit away from the wall if they flew in.

4.3d Roof

The roof can be either a completely solid structure made of plywood (that is painted for weather protection or covered with roof shingles), fiberglass, or a combination of solid and nonsolid materials (welded or heavy duty plastic wire, vinyl coated chain link, wooden slats, or dowels). A solid covering should extend at least a third the distance from the back of the cage forward.

However, if your area experiences heavy snowfall, it is recommended to have at least one-half of the roof solid. This will provide an area where your bird will find protection from the sun, rain, snow, and other elements. The forward portion can be made of wire (plastic-coated or welded), wooden slats, or dowels, to provide fresh air and sun. However, TRC recommends constructing any nonsolid roof portion out of a double layer of material to protect your bird from other aerial predators. Any solid sections of the roof should be sloped to allow proper drainage of rain and snow or constructed of a heavy enough material to withstand your maximum rain or snow load.

Also, as mentioned for nonsolid walls, it is recommended to

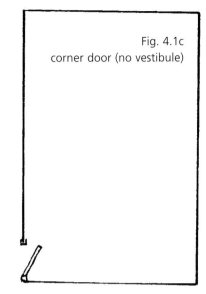

Fig. 4.1c
corner door (no vestibule)

Fig. 4.1d
Portable vestibule.
(Reprinted with permission,
Journal of Wildlife Rehabilitation
2003, Vol 26, 1: 23.)

apply mosquito netting on the outside to help protect your bird from mosquito transmitted diseases (chapter 7, Medical Care).

4.3e Floor

The floor must provide quick drainage of water, be easy to clean, and not be abrasive to your raptor's feet. Floors covered with pea gravel, grass, or artificial turf all work very well. Sand can also be used, but birds have been reported to ingest sand with their dinner and develop crop impactions (chapter 7, Medical Care). TRC does not recommend using wood chips, hay, straw, or uncovered concrete. Wood chips, hay, and straw hold moisture and provide an ideal environment for mold, and concrete is abrasive to raptor feet.

If you decide to use gravel, use only rounded buck shot gravel (¼ in/0.6 cm). Larger rock and gravel with sharper edges can lead to serious foot problems. In addition, gravel (and sand) substrates should be built up to at least 4 inches (10.2 cm) in thickness.

4.4 PREDATOR CONTROL

When constructing the enclosure, another important consideration is predator control. There are a couple of approaches you can take in construction of the facilities to help prevent predators from digging their way into the cage.

First, the bottom of the enclosure can be made with a concrete base or a wire lining. Concrete should be angled down to improve drainage out of the enclosure and both must be covered with one of the substrates recommended above for drainage and foot care.

A second option is to place 1 x 1 inch (2.5 cm x 2.5 cm) or 1 x 2 inch (2.5 cm x 5.1 cm) welded galvanized wire 2–3 feet (0.6-0.9 m) underground, at a 45-degree angle away from the cage. Hot wire placed around the lower portion of the mew is also an option to help prevent potential predators from entering.

However, even with one of these features in your enclosure design, if you think a mink, raccoon or other predator is visiting the area, place an un-baited live trap along one wall of the cage every night. As the mammal is searching for a way in, it may enter the trap and can then be removed from the area. You can never be too protective.

One last note on cage construction: be sure that all the seams and connections between materials are secure and tight. Don't allow any nails, screws, or wire to extend into the cage. Any protrusions are likely to injure your raptor. Remember, when it comes to accidents, injury, or escape, if it can happen, it will!

4.5 FURNITURE

A house is not a home until you have a place to sleep, a place to eat and drink, and a comfortable place to sit and relax. This is as true for your raptor's home as it is for yours. Thus, your raptor's mew needs to contain the proper furniture: a hutch or shelter box, one or more perches, and a water pan.

4.5a Hutch

A hutch (figure 4.2) or shelter box (figure 4.3 a,b) will provide additional protection from inclement weather, as well as give your bird a place where it feels safe from predators. Both of these structures can be made from plywood and, if protected with a water sealant, will last for years.

For sanitation purposes, the shelter box should have an open bottom, and in the hutch shown, perches can be placed both inside

Fig. 4.2
A bald eagle tethered to a hutch.
(Gail Buhl)

Fig. 4.3
Shelter boxes.

(a)

(b)

and on top (tethered birds often like to perch on the top of hutches). Some caretakers prefer an A-frame hutch (just as it sounds, the frame is shaped like the letter "A" and is pointed on top), which discourages a bird from perching on top.

Recommended hutch and shelter box sizes are listed in table 4.2.

4.5b Perches

Providing your raptor with appropriate perches is one of the most important things you can do to keep it comfortable and healthy. Perches need to be individually customized in shape, design, and placement for each bird. It is important to offer your bird a variety of perches surfaces (table 4.3). Improper perches will result in bumblefoot, an extremely serious foot condition (chapter 7, Medical Care).

Perch placement

Generally, raptors care less about what they perch on than where they perch. This means that for free-lofted birds you need to put perches in the spots where your raptor wants to stand. And since most birds will use more than one spot, often varying with the time of day and/or the season, you will need more than one perch. Many hawk species like to take advantage of the morning sun, while many owls are more comfortable in darker settings. It will be helpful to do a little investigating and find out behavior patterns of the species you are going to house.

Another consideration when placing perches is retrieval of your bird. Make sure that the perches are placed at a height that will permit you to easily get your bird on your gloved hand (if manned) or grab it (if unmanned and in a display). You also want to place perches such that your bird can comfortably open its wings and turn around without hitting the walls, roof, or floor of the enclosure. Adequate distance from these objects will depend on the bird's size. In addition, it is essential to place at least one perch under a solid part of the enclosure's roof to offer the bird protection from the elements.

Once you decide on the locations to place perches, you will need to decide how to secure them. For a free-loft mew, perches are most often either secured solidly to the sides of the mew or suspended from the roof with rope. Wall perches are often placed at an angle in a corner or span the entire width of the enclosure and are secured to two walls. However, sometimes, free-lofted raptors choose to perch close to the wall and when they turn around or open their wings, their tail and primary flight feathers rub against the wall resulting in feather damage. An innovative perch design created by John Karger, Last

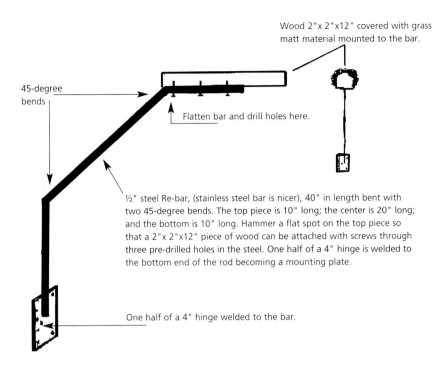

Wood 2"x 2"x12" covered with grass matt material mounted to the bar.

45-degree bends

Flatten bar and drill holes here.

½" steel Re-bar, (stainless steel bar is nicer), 40" in length bent with two 45-degree bends. The top piece is 10" long; the center is 20" long; and the bottom is 10" long. Hammer a flat spot on the top piece so that a 2"x 2"x12" piece of wood can be attached with screws through three pre-drilled holes in the steel. One half of a 4" hinge is welded to the bottom end of the rod becoming a mounting plate.

One half of a 4" hinge welded to the bar.

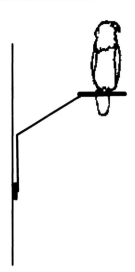

Fig. 4.4
Wall perch mount. (Design by John Karger, Last Chance Forever/The Bird of Prey Conservancy)

Chance Forever/The Bird of Prey Conservancy, suspends a perch from a sidewall by a special mounting bar that provides feather clearance when a bird repositions itself (figure 4.4).

Suspended perches more closely resemble tree branches because they move a little with the bird as it lands (figure 4.5). If you choose to place perches in this fashion, make sure to suspend them with strong material. Avoid thin wires or cables, as the birds can "miss" these and fly into them. For tethered birds, a perch should either be staked into the ground or constructed with a heavy base to prevent shifting.

Fig. 4.5
A natural branch perch suspended from the roof of a mew. (Gail Buhl)

Perch design

Regardless of whether you free-loft your bird or tether it, there are two basic styles of appropriate perches: those with flat surfaces and those with rounded or beveled surfaces.

Flat perches: Falcons and ground dwelling species most commonly use a flat perching surface. These birds need flat surfaces to perch on because of the special design of their feet. Forcing them to perch on rounded perches will undoubtedly lead to bumblefoot. For free-lofted birds, natural flat perches such as large rocks, tree stumps, and wooden fence posts are often used as well as man-made platform (shelf) perches (figure 4.6). Platform perches are usually made from wood and covered with long-leaf artificial turf (Monsanto Industries), coco mat, or a thick indoor/outdoor carpeting (stadium matting).

Catalogs that include these products are listed in appendix D. Ready-made block perches for tethering falcons, vultures, and ground dwelling species are also available from a variety of the suppliers listed.

Rounded perches: The other style of perch is rounded. The trick with rounded perches is to provide your bird a variety of diameters and surfaces. If your bird is free-lofted, multiple perches can be placed in the mew to provide this variability. Rounded perches can be specifically designed out of wood for different species (figure 4.7) and then covered with untreated manila three-strand rope or artificial turf.

Natural branch perches also can work well as long as a few perches of varying diameters are provided, sharp protrusions are removed, and the perches are replaced regularly when the bark

Fig. 4.6
A platform (shelf) perch commonly used in the mews of large falcons. (Gail Buhl)

Fig. 4.7
Bevel cross-sections for specialized raptor perches. (Jen Veith)

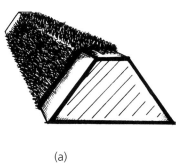

(a)
Eagles

(b)
Red-tailed hawk, great horned owl

(c)
Rough-legged hawk

(d)
Barred owl

gets smooth. The use of worn out natural perches can lead to severe foot problems.

Another rounded perch option is plastic conduit (or pvc pipe) wrapped with rope or artificial turf. However, due to the uniform roundness of these perches, several perches of varying diameter must be provided and a close watch kept on the condition of your bird's feet.

A last type of rounded perch sometimes used has a rod iron base. A black flexible hose is wrapped around the iron rod by making one slit down the length of the hose. The hose is then covered with rope or Astroturf. The hose provides a little cushion and increases the diameter of the perch (since rod iron often comes in small diameters). However, you must be careful not to end up with a perch that has too large a diameter for the species you are housing.

When choosing rounded perches for your free-lofted bird, there are several things to keep in mind. First, raptor species differ in the size of their footpads, the distance between toes, and the weight that must be borne on the feet. These factors must be considered when perches are constructed out of wood. To mold an appropriate perch for your raptor's feet, specific angles can be cut to prevent bumble-foot (figure 4.7).

Second, it is very easy to fall into the trap of providing your bird with perches that are too large in diameter. Perches that are too large cause severe problems with the footpads and rear toes. If you haven't had the opportunity to see raptors in the wild, go on a fun bird-watching trip. You will notice that raptors often perch on branches with small diameters.

Third, the longevity of the materials you use will depend on your climate. Rope, wood, and natural perches rot more quickly in areas with a lot of moisture and/or high humidity and will need to be replaced more often.

Lastly, feel free to mix and match different types of rounded perches in the enclosure. The more variety you can offer, the healthier your bird's feet will be.

If your bird is tethered, you can either rotate its perch every few days to provide variation in diameter and surface, or provide a second perch that the bird can reach but not get tangled in or around. Many hawk and eagle species are often tethered to a hutch with a bow perch provided outside the hutch. Birds tethered with this type of arrangement should be able to reach the second perch, but not go beyond it. Ready-made bow perches for tethering are available from sources listed in appendix D. Tables 4.3 and 4.4 list recommended perch designs for free-lofted and tethered raptor species, respectively.

Perch height

When designing perches for your bird, it is important to not only consider the shape, but also the height for non-flighted or tethered individuals. Different species have different tail lengths and perches need to be high enough off the ground to provide tail clearance, but not so high that disabled individuals cannot reach them or could plummet a long distance to the ground and hurt themselves.

Perch maintenance

As mentioned above, the surfaces of perches wear in time and must be replaced to prevent foot problems. Artificial turf becomes thin and worn, natural branches lose their rough bark and become smooth, and rope becomes smooth, rotted or frayed. When perches reach these conditions, they must be replaced. Natural branches can be entirely removed and replaced with new ones. Wood or PVC perches covered with artificial turf or rope can be reused (if the wood is not rotted or moldy) provided the covering is replaced.

To replace artificial turf, coco mat, or stadium matting on block perches, all-purpose glue, such as contact cement, can be used. To replace the perch covering on a rounded perch, tightly wrap it around the perch and fasten it. Horseshoe nails can be used on the end of wood perches to secure rope, and horseshoe or regular nails to secure turf or matting. To secure rope or matting to PVC perches, hose clamps can be used provided they are covered with a thick tape, such as electrical tape or duct tape, to cover rough, sharp edges. Rope can be secured on PVC or rod iron perches by placing several inches of rope horizontally along the perch end and wrapping the rope tightly in a vertical direction over the PVC.

To replace rope, wind it tightly around the perch, making sure that each wrap is pushed snugly next to the previous wrap. For added perch texture, a perch can be wrapped with two layers of rope, the first layer horizontally tight as just mentioned and then the second layer wrapped such that a space equal to the diameter of the rope exists between each wrap. However, if a double layer is used, you must make sure that when you are done, the perch diameter has not gotten too large for the species you are housing (you may have to start with a smaller diameter perch).

4.5c Water Pans

Most raptors drink water or bathe, particularly during warm weather. Drinking water can be provided in appropriately sized dog bowls. Bathing water can be provided in pans or pools.

For small raptors, such as American kestrels, sharp-shinned hawks, merlins, Northern saw-whet owls, and Eastern and Western screech owls, you can offer bathing water in 8-inch (20.3 cm) diameter plastic dog dishes (with rounded edges).

For medium and large birds (long eared owls, Mississippi kites, great horned owls, Harris's hawks etc.), use medium-sized plastic pans (18 in x 25 in x 7 in [45.7 cm x 63.5 cm x 17.8 cm] with rounded edges), or similar-sized rubber pans.

For extra large birds (eagles), use large plastic pans (3 ft x 2 ft x 8 in [0.9 m x 0.6 m x 20.3 cm]), plastic children's swimming pools, or specially made structures buried into the ground (for free-lofted birds only).

The primary rule is to keep the water shallow enough so the raptor cannot drown (not having water extend any higher than the bird's belly is a good rule of thumb) and the top of the pan should not extend above the bird's head (birds are often not comfortable going into deep structures and when wet may have a difficult time exiting a deep structure).

Be especially careful if your bird is missing any part of a wing or has very limited extension of one wing. Birds with these conditions often have a difficult time righting themselves and can easily drown if the water is too deep. They should be offered water no deeper than 1–2 inches (2.5–5.1 cm).

If your bird is tethered, the water pan should be placed at a distance that allows the bird to reach the middle of the pan, but not cross over to the other side. Also, keep an eye on where your bird perches. If it routinely perches on the edge of the water pan, move the pan to a different spot and offer a perch where the pan was. The edge of a water pan is not a good surface for your bird's feet and if the bird perches with its tail constantly in the water, the tips of the tail feathers can rot off.

Also, keep in mind that you want to place water pans such that they will not be a reservoir for bird mutes. If a bird is tethered, place the pan to one side of the perch. If your bird is free-lofted, place the perch in an open area away from where mutes will reach. Do not place water directly below, behind, or in front of a perch.

For several reasons, TRC does not recommend providing your bird with bath or drinking water in freezing temperatures. First, most birds won't bathe in extremely cold weather and if they would, ice build-up on their feathers could prevent them from properly keeping themselves warm. Second, a bird's equipment (jesses, leash, swivel) can get icy in bath water. This could cause frostbite to a foot/leg area that is in contact with anklets or jesses. Also, for tethered birds, swivels can get iced and malfunction resulting in an increased possibility that your bird could get twisted. If you want your bird to have the opportunity to bathe or drink when outdoor temperatures are at or below freezing, you can bring it into a warmer environment (above freezing temperatures but not hot), offer it a bath, and allow it and its equipment to completely dry before replacing it in its mew.

Water should be changed frequently, to prevent the build-up of algae, bacteria, and mosquito breeding. Cleaning water pans a minimum of every other day should help avoid these situations. Disinfecting water pans twice a week should suffice (see 4.9 Sanitation).

4.6 CUSTOMIZING ENCLOSURES

4.6a Visual Separation Between Cages

If birds of different sizes or species are kept near each other, it's a good idea to offer them an area where they are visually separated. Some species feel threatened by the close proximity of other species (for example, screech owls and barred owls, peregrine falcons and great horned owls), which will cause them to become anxious, flighty, and possibly stop eating.

The easiest way to separate species is to put a solid wall between two cages. This will allow you to keep your birds next to each other for easy access, or even to use a common solid wall for two cages. If you separate species by a wall that is not completely solid, the nonsolid portion must have two layers of mesh or wire separated by two inches (5.1 cm) of space; otherwise, one bird can reach through and grab the bird in the adjacent enclosure.

4.6b Options for Shelter Boxes

As an alternative or an addition to the hutch and shelter boxes pictured in figures 4.2 and 4.3, nest boxes will be used by many species. American kestrels, barred owls, common barn owls, and the smaller owls in particular will roost in these boxes. The advantage of using a nest box is that it might give your raptor an extra feeling of security. The disadvantages are that the bird will be hard to see if on display, may be hard to get out of the box, and the box might encourage nesting behavior, particularly in human imprints.

Designs for nest boxes can be found in Caroll Henderson's *Woodworking for Wildlife: Homes for Birds and Mammals* (see suggested readings at the end of this chapter) and on the web.

4.6c Modifications for Physically Disabled Raptors

Birds that are limited in their flight ability may require special adaptations in their housing arrangement. Many species want to be perched on the highest perch possible and birds with limited flight may have difficulty reaching perches from the ground. A series of perches connected to each other at 45-degree angles will let a bird walk from the ground up to a perch. However, TRC recommends this arrangement only for birds that have at least some use of both wings and have the ability to controllably glide to the ground if they jump off. Birds that cannot fly or glide should only be offered perches close to the ground. If these individuals are offered access to higher perches, they could potentially sustain a serious injury or even die from jumping off and plummeting to the ground.

4.6d Housing More Than One Bird in an Enclosure

Although it's possible to house free-lofted birds together for display, you must be extremely careful in deciding which birds to house together, and you must watch them closely. Keep in mind that raptors are carnivores, some raptor species prey on other raptors, and some are primarily bird eaters. Here are some guidelines to follow:

- Most individuals of the same species can be safely housed together. Exceptions include human imprints (once they reach sexual maturity), accipiters of the opposite sex, and small falcons (kestrels, merlins) of the opposite sex.

- Do not mix species of different sizes (exception: TRC has successfully housed osprey and turkey vultures with smaller hawk species such as broad-winged hawks and red-shouldered hawks).

- Be careful when adding a bird to an existing display. Often, the long-term resident will become territorial of its mew and work to drive the new bird out. This behavior is most common during the breeding season.

- A few species do especially well when housed with other members of their own species. These include, eastern and western screech owls, northern saw-whet owls, bald eagles, golden eagles, osprey, Mississippi kites, and swallow-tailed kites.

- Bald eagles and golden eagles often do not mix well and TRC houses them separately.

- Great horned owls are a threat to almost any species and TRC recommends always housing them separately.

- Burrowing owls develop strong territorial behaviors when housed in pairs and should not be housed with other species of small birds (they may kill them).

- Some species can be housed together. Table 4.5 lists combinations of species known to be successfully housed together. This information is based on TRC's direct experience and communication with other educators. Other combinations of species may also work, but TRC recommends contacting local rehabilitators, educators, or other raptor caretakers to get their input. The size and shape of the enclosure, along with the personality, age, disability, and natural history of the individual birds all play a role in compatibility.

Keep in mind that every bird is an individual. Even though people have successfully housed individuals listed in table 4.5 together, this may not mean that all individuals of these species will express the compatibilities listed. Thus, it is critical to observe birds closely once you put them together. Monitor their weights regularly to make sure everyone is getting their fair share of food. If any signs of aggression are present (raised hackles, chasing, vocalizing, posturing, fighting), separate the birds immediately.

If you are going to house birds together, the size of the enclosure must be larger than those listed in table 4.1. The adequate size will depend on the number of birds and the species involved (if you are mixing a few). The larger the enclosure, the fewer negative interactions the birds will have with each other. It is highly recommended to do a little research and find out from experienced caretakers what size enclosures have worked well for them.

In addition, it is critical to provide an adequate amount of food (in separate piles for each bird), perches, and shelter boxes.

Choose birds that appear even-tempered, not territorial or aggressive. TRC recommends slowly introducing birds to each other. This can be done by temporarily housing a new resident in an adjacent mew with visual access so the two birds can get accustomed to each other. Also, when putting birds together, it is a good idea to move both birds into a totally different enclosure. This will help reduce territorial aggression.

As mentioned earlier, no matter which birds you put together, observe them closely to make sure they are compatible. Don't try to make the "most creative" display. Weigh the birds regularly, and as long as they are housed together, monitor their behavior and general physical appearance (condition of feathers, feet, and wrists) for any changes that could indicate a problem. Unknown "triggers" can result in conflict and aggression between birds that have been housed together peacefully for a long time.

Due to the unpredictable behavior of raptors in captivity, the safest bet is to house birds individually. If you must house more than one bird together, however, and you are unsure about mixing specific individuals, contact rehabilitators, falconers, zoos, or other bird caretakers for their input.

4.6e Supplemental Heat

As indicated in chapter 2, Selecting a Bird for Education, several species of raptors are extremely sensitive to cold temperatures and have difficulty keeping their appendages and core body temperatures warm. If you are housing one of these birds in a cold climate, you must provide it with extra warmth during the cold season. There are several ways to accomplish this.

First, you can insulate smaller mews with a layer of thick plastic applied to the outside walls and roof (if open and slanted to allow snow removal). This decreases the effect of wind chill and helps retain solar warmth provided during sunny days.

Second, you can provide a supplemental heat source to sensitive species (small falcons, migratory species). This might allow you to keep your bird outside even when the weather gets really cold. There are several heat options that have been proven safe. The most important considerations with all sources of supplemental heat are that the bird cannot directly touch the heat source and that the mew temperature is monitored to prevent overheating.

Light bulb heater unit (figure 4.8a)

For free-lofted species that are cavity nesters, additional heat can be provided by modifying their nest box to include a second chamber containing a heat source. The main nest box should be accessible to the bird and have at least three solid sides and a solid roof. Then, a second chamber about one-half the height of the nest box can be attached to one side of the box and contain two heat generating bulbs. Recommended bulbs are 60-watt Longer Life Bug-A-Way bulbs placed into ceramic sockets and oriented to face toward the main box. The bulb chamber should have a solid roof and outer walls, as well as a removable floor for easy bulb removal. Small ¼–½ inch (0.6–1.3 cm) diameter air holes should be created between the small and large chambers to allow heat to easily pass into the nest box. This arrangement typically maintains the temperature in the nest box at about 10°F (−12.2°C) above the ambient temperature.

Fig. 4.8 Heater options.

Fig. 4.8a
Light bulb heater unit in attached enclosure. (Ron Winch)

Fig. 4.8b
Reptile heater. (Gail Buhl)

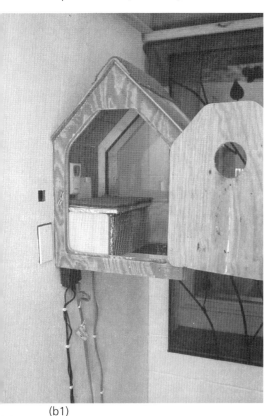

(b1)

(b2)

Reptile heater (figure 4.8b)

A heater system designed for reptiles can also be modified to keep a hutch/shelter box warm. One system used is called ZooMed® Heat Control (www.zoomed.com). With this system, the heating element should be placed in a fireproof, waterproof box within the hutch and a wireless thermometer within 1–2 inches (2.5-5.1 cm) of the heat source. This arrangement allows the temperature in the hutch to be closely monitored. Different voltage/wattage ZooMed ceramic heaters can be used depending on the temperature; 50–100 watt bulbs work really well, with the higher watt bulbs used during the coldest weather. Ceramic bulbs are especially nice because they provide heat but do not give off any light.

Space heater

A space heater placed in a small mew can be effective in providing extra warmth to your bird during cold days. However, in order for this to work, the mew must have a solid roof and open sides must be covered with plexiglass or heavy duty plastic so the heat does not immediately escape outside. In addition, as with any electrical device, you must make sure the bird cannot touch the unit (or electrical cord) and that the unit is protected from moisture. Also, the heating element must not be coated with Teflon, as vapors emitted from heated Teflon are toxic to birds.

New products

It is always recommended to check with peers in the field to find out about new products. Wolf Ridge Environmental Learning Center in Finland, Minnesota is currently working with another company, Helix Controls, to design a thermostat specifically for outdoor bird hutch use. This system will have a circuit that automatically shuts off if an animal bites through the cord and a default temperature that will keep the temperature in the hutch at 32°F (0°C). The system will have proportional warming and cooling so it would have slow warming and slow cooling ability. For more information, contact Helix Controls, 1035 East Vista Way, Suite 150, Vista, CA 92084, 720-726-4464, helixcontrols.com. In addition, the World Bird Sanctuary in St. Louis, Missouri (www.worldbird sanctuary.org) has designed a safe system for heating perches that should be patented soon.

Even if you choose to furnish your bird's mew with a supplemental heat source, it is a good idea to bring a sensitive species/individual into a slightly warmer environment such as a garage or unheated building on really brutal days or nights. However, do not bring it into a 70°F (21.1°C) structure, as it will have a harder time adapting to the wide range of temperatures when put back outside. (If you live in a climate with great seasonal variation, you have experienced this first hand. Forty degrees Fahrenheit (4.4°C) in November feels cold and you need an extra jacket. In March, this same temperature feels warm and some people actually take jackets off.)

4.7 TETHERING

When you are designing a housing enclosure that meets the standards listed above, you should consider whether you will tether or free-loft your bird. If your design is for a small mew, tethering the bird might require a slightly larger space. The size and design of the mew and the placement of a hutch, shelter box, and perches will depend upon the way in which your bird is managed.

Tethering is the process of restraining your bird by tying it to a perch (figure 4.9, 5.1). The perch can be freestanding, secured to the ground with a spike (figure 4.10), attached to a hutch (figure 4.2), or secured in a stall (figure 4.11).

Why are birds tethered? Can any raptor be tethered? What equipment is needed? What precautions must be taken when a bird is tethered? What is the proper way to tether a bird? There are numerous things to consider about tethering your bird.

Fig. 4.9
A great horned owl properly tethered to a bow perch. (Gail Buhl)

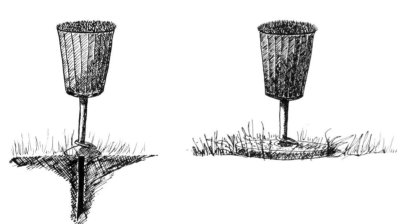

4.10
Perches most commonly used for tethering raptors.

(a1) (left)
Block perches — spike.

(a2) (right)
Block perches — base. (Gail Buhl)

(b1) (left)
Bow perches — spike.

(b2) (right)
Bow perches — base. (Jen Veith)

4.7a Why Tether a Bird?

Raptors are tethered for many reasons: to facilitate training, to prevent a bird from injuring itself in its mew, to restrain a bird in a weathering area for part of a day, or to restrain it at a display site.

Training

As mentioned in chapter 8, Training, handling a new bird is much easier if it is tethered instead of free-lofted. Tethering restricts the bird's movements and causes it to face stressful things rather than hide from them. Repeatedly exposing a bird to an intimidating situation will decrease its sensitivity to it.

Part of the training process is to establish a routine that lets your bird know what is going to happen when you approach and what behavior you expect from it. Ideally, retrieving a trained raptor should be as simple as entering the mew, putting your glove out and "asking" your bird to step up. If you free-loft your bird (especially during the training process), you give it the freedom to move around and fly away from you. It takes much longer to give your bird the message that the appropriate behavior is to stand still while you approach and then step onto a glove. You need to be in charge and not give your bird behavior options to choose from.

Acclimation

Tethering is also a safer way to house a bird during acclimation to a new housing arrangement. It will prevent the bird from flying away from you in a frenzy and possibly grasping the sides or roof of the mew or crashing into the perches or shelter box. Also, the handler will be safer because a tethered bird is under better control, decreasing the likelihood of injury to the handler.

Safety

Some birds are too energetic to handle a free-lofted situation well. These birds should be tethered in their mews so they don't injure themselves, break feathers, or injure handlers.

Weathering

Often, a bird might need to be housed in an enclosed environment. Such a bird should be transferred for part of the day to a more open, sunny area so it can "catch some rays" and bathe. This process is called weathering. If the weathering area lacks a roof and sides and/or a vestibule, as most do, the bird must be tethered to a perch to prevent escape.

Display

Finally, many bird handlers periodically put their birds on display outside the mew. If a bird is not held on the fist during such a display, it is usually tethered in an observation area and constantly supervised.

4.7b Which Birds Should Not Be Tethered?

There are a few situations in which a bird should not be tethered. First, if a bird has only one functional wing (either a full-wing amputee or a bird with loss of complete wing function due to nerve or muscle damage from the shoulder to the wing tip), it should not be tied to a perch. These birds often have balance problems and can easily get tangled in a leash. They need to have unrestricted movement to maintain balance. Also, if they are tethered to a perch and jump off, there is a good likelihood that they'll have difficulty getting back on, or may fall over and not be able to right themselves. Obstacles, such as ramps, should not be used in a tethering situation because a leash can easily wrap around them.

Second, birds with leg problems, such as poorly aligned healed fractures or arthritis, should not be tethered. If such a bird bates and constantly hits the end of the tether, its sensitive legs could become injured or the arthritis aggravated.

4.7c Can Tethered Birds Be Housed Together?

Safety should always be your number one concern. For a raptor's primary housing, the best and safest situation is to house tethered birds separately. If this is absolutely not possible, here are a few rules to follow:

- Species must be compatible (table 4.4). If one bird breaks its equipment and gets loose in the enclosure, you don't want it to prey on other birds.

- There must be a visual separation/barrier between birds. This prevents birds from bating toward each other or exhibiting other undesirable behaviors such as mantling over food, vocalizing, etc.

- The birds should never see each other eat in the enclosure, as this can cause excessive bating and aggressive behaviors.

Although TRC does not recommend tethering multiple birds together in a primary housing facility, many different species can be tethered together for limited periods of time in a weathering area. However, they must have ample room to reach the end of the leash without grabbing another bird, no birds should be fed while in this area, and the area must be constantly monitored by a skilled caretaker so he/she can intercede if undesirable behaviors or equipment malfunction occur. Many education facilities safely display raptors in this fashion during public events.

4.7d Equipment

If you decide to tether your bird, you'll need several basic pieces of equipment: a pair of jesses, a swivel, a leash, a secured perch, and a ring to tie the leash to. Chapter 5, Equipment, offers detailed information on jesses, swivels, and leashes. Appropriate perch types and heights are listed in table 4.4.

If your bird has a tendency to get twisted while tethered, there are a few equipment modifications you can make to help prevent it.

Jess extender
A jess extender is a 6–8-inch long (15.2 -20.3 cm) piece of leather or nylon cord with a slit or loop at each end (figure 5.19 and 5.20) that is placed between the jesses and swivel. It is used to relieve tension if your bird turns around frequently and prevents the swivel from looping between the jesses. More information on jess extenders and how to make them is located in chapter 5, Equipment.

Double swivel system
For small and medium-sized birds, a double swivel system is an effective way of preventing jess twisting. This system involves three swivels, one attached to each jess and both of them attached to the third swivel (figure 5.4). This provides additional rotation to release tension and prevent jess twisting. The negative side to this system is that it requires your bird to wear additional "hardware" which is often undesirable. The double swivel system is described in more detail in chapter 5, Equipment.

4.7e Tethering Stall

Sometimes, it is handy to have a temporary indoor housing arrangement to tether a raptor that is in training and/or needs a place to reside during temperature extremes or severe weather events. One type of arrangement that works well is a "stall" consisting of three solid sides with a perch. Depending on the species, a shelf perch, bow perch, or 2 x 4 inch (5.1 cm x 10.2 cm) perch system securely attached can be used (figure 4.10).

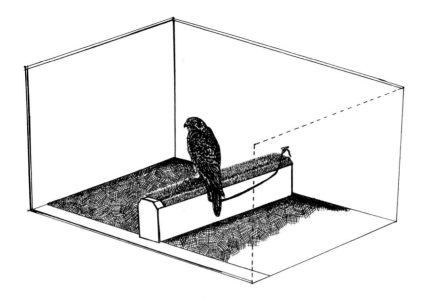

Fig. 4.11
Tethering stall. (Jen Vieth)

4.7f Precautions: Tethering Dos and Don'ts

Whenever you tether your bird, there are several things you should consider to ensure your bird's safety.

- Make sure you choose an appropriate type of perch that is safe. The perch should not allow feather breakage or tangling of the leash. Recommended perch types and heights for tethering are listed in table 4.4.

- The bird should not be tethered to the same perch all the time. Either rotate perches every few days, or offer a second perch the bird can jump to but not over.

- Make sure the perch is placed in a safe area, free of obstacles. Your bird should always have sufficient room to bate without hitting anything. The leash should be long enough to allow the bird to easily clear the perch if it bates; to reach the ground, unless it is tethered to a shelf perch; to reach its water, stand in it, and bathe, but not pass over it (otherwise its leash could get stuck on the bath container); and to fly up to the top of a hutch, if tethered to such a hutch. It should not be so long that the bird can reach undesirable areas, the end of a shelf perch, another bird tethered nearby, or over the back of the hutch.

- Do not tether your bird to a perch placed on a table, unless the perch is so heavy that the bird cannot move it much if it bates or the perch is secured to the table. Also, the bird must be tethered so that it cannot reach over any sides of the table and potentially hang.

- Place the perch on a soft grass or sand surface. Don't place the perch on cement, tile, or any other hard surface. If your bird bates onto such a hard surface, it can easily damage its feet, wrists, and feathers.

- Don't tether your bird in a mew it was just free-lofted in. The bird will be used to reaching high perches and will repeatedly try, not understanding its new restrictions.

- If your bird is tethered in its mew, don't add high perches in anticipation of free-lofting it someday. This will make your bird extremely agitated as it tries to reach the more desirable high perches. You can add high perches when and if you decide to free-loft the bird.
- If tethering your bird outside for weathering or display, make sure it has access to sun and shade. Birds can overheat rapidly.

- If tethering your bird outside in an area that is not completely enclosed, someone should be present to monitor it at all times. Predators, such as dogs, cats, humans, aerial predators, and even bears are a potential risk to tethered, unprotected birds. Never leave food for a bird tethered in an area that is not completely enclosed. This attracts other predators (aerial and ground) and will put your bird in danger.

- Do not tether your bird outside and/or near other tethered birds with a jess extender on (chapter 5, Equipment). A jess extender has only one secure point of attachment to the swivel and if it breaks, your bird is free. (Without the extender, each jess is attached to the swivel so there are two straps that secure your bird).

- Frequently monitor all tethered birds. Sometimes a bird becomes active on its tether and repeatedly turns around on its perch, until its jesses become very twisted. This can decrease circulation to its legs and feet, causing severe and potentially permanent injury. In freezing temperatures, a bird's feet can become frostbitten in a very short time if the bird is twisted and not able to keep its feet warm. A tethered bird must be checked several times a day to ensure that this has not occurred.

4.7g The Proper Way to Tether a Bird

Now that it has been established which birds should be tethered and the safety measures that should be taken when doing so, let's focus on how to initially get a brand new program bird tethered to a perch. Unless your bird has been previously trained, a new program raptor is not going to perch on your gloved hand and allow you to transfer it this way to its tethering area. So, how do you tether it in the first place?

Many times a new, previously wild, bird will need to be grabbed and carried in a cradle position (chapter 6, Maintenance Care) to its housing enclosure. Then, to tether it to a perch you will need to either clip the leash to the perch ring or tie a falconer's knot (figure 4.12). If you are by yourself, you will need to rearrange your hold such that you secure the bird with one hand/arm and free up your other hand to tie the knot. Otherwise, you can have an assistant tie the knot. With a little practice ahead of time, a falconer's knot can be easily tied with one hand. It is a very secure knot and can be untied with one hand as well. TRC prefers using falconer's knots to clips that can become faulty.

Once the leash is secured to the perch, TRC recommends gently placing the bird on its side on the ground, and gently rolling it away from you as you remove your hold. This will offer the bird a few seconds to stand and get its bearings while you retreat. If you place it on its feet or attempt to place it directly on the perch, it will probably try to fly away from you immediately and sharply hit the end of the tether. It is a more positive and comfortable approach to let a new bird "find" the perch itself.

Once the bird is safely tethered, all subsequent handling should be part of a training program as described in chapter 8, Training.

4.8 ENRICHMENT

In the wild, raptors are continually stimulated by their surroundings, their ability to perform natural behaviors such as hunting, breeding, and migrating, and the interactions they have with other living things. In captivity, their entire world changes and a lack of social and environmental stimuli can lead to behavioral problems such as feather plucking (especially in human imprints), pacing, aggression, and stress-related medical problems. As a raptor caretaker, it is your responsibility to not only maintain your bird's physical health, but also its mental health by providing it with things that will enrich its daily life.

In 1998, David Shepherdson described enrichment as "an animal behavior principle that seeks to enhance the quality of captive

Fig. 4.12 The proper method for tying a one-handed falconer's knot. (Gail Buhl)

(1)
With the leash passed through the ring on a perch, slip your index finger between the two strands from underneath.

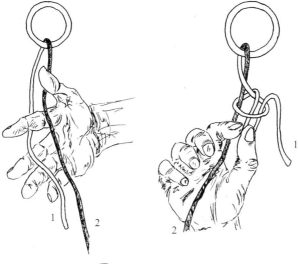

(4)
Slide strand 1 between your second and third fingers, through the loop formed by your thumb. Pull tight.

(2)
Slip your thumb over strand 2 and under strand 1 and pull strand 1 across strand 2.

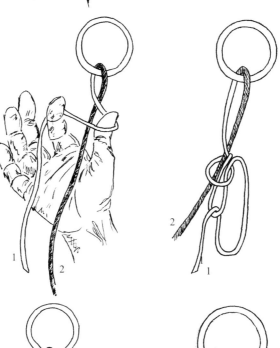

(5)
Take the end of strand 1 and pass it through the loop.

(3)
Bring your index and third fingers under strand 2 and above your thumb.

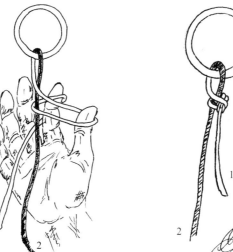

(6)
Pull strand 2 to tighten the knot.

animal care by identifying and providing the environmental stimuli necessary for optimal psychological and physiological well-being." For a raptor, this often involves providing things that are visually stimulating, offers it natural choices, and allows it to perform natural behaviors. Keep this in mind when designing a housing structure.

First, pick a location that provides a natural view; do not tuck your bird between buildings or in another visually limiting situation. Next, a raptor should have sufficient space to move around, eat, bathe, fly (if flighted), "hide" (nest/shelter box), and have a choice between perches. The amount of space a bird should be offered will depend on its size, injury/flight ability, the way it is housed (free-lofted vs. tethered), and the number of individuals housed together.

In addition to housing location, size, and furniture, there are other things that have been provided to raptors for enrichment. Here are a few enrichment ideas for program and display raptors:

- corncob
- dog chew toy
- dog Kong toy
- handling/training
- live food (if deemed safe with your bird's handicap and housing arrangement)
- phone book
- tennis ball
- varied diet (necessary for good health anyway)

When choosing forms of enrichment for your captive raptor, keep in mind the safety of your bird, the natural history of the species, personality of the individual, and the cost and ease of integration. Remember only you know what is safe for your bird. The results of a good enrichment program will include a healthier animal physically and mentally, decreased aggression and stress, and increased immune health.

For more information about enrichment, please visit the IAATE website: www.iaate.org and see the suggested readings at the end of this chapter.

4.9 SANITATION

"Wash your hands before supper!"

How many times did we hear this while growing up (and maybe even now)? There were good reasons why our parents made us clean up before eating and why good personal hygiene is so important for us, and for the animals we work with. Bacteria, fungi, parasites, and viruses can easily enter our bodies when we use contaminated hands to eat food. Good personal hygiene can help reduce the possibility of infection or spreading disease to other people and animals.

As you can imagine, it's much easier to explain the importance of good hygiene to people than to animals. The raptor in your care isn't going to understand that if it stands on its mutes, it shouldn't hold food in its feet when eating. As a raptor caretaker, it's your responsibility to create a hygiene plan for your bird that includes regular disinfecting of its housing enclosure.

4.9a Disinfection

When we talk about disinfection and raptor hygiene, we're referring to destroying pathogenic organisms in your raptor's residence area, providing it with clean water for drinking and bathing, and offering it only food that has been stored properly. Here are some guidelines:

- Design your bird's mew so that it is easy to keep clean and disinfected. Use easy-to-clean materials in its construction, and place perches so that mutes will fall on the ground, not on the walls, feeding platform, or water dishes.

- Practice good personal hygiene by wearing rubber gloves when picking up old food and cleaning mews.

- Remove uneaten food within twenty-four hours. In the summertime, bacteria, flies, and maggots flourish and multiply in the leftovers. Also, food remains may attract bees, hornets, yellow jackets, and predators (such as weasels, raccoons, opossums, and skunks).

- Disinfect your bird's mew regularly. Keep in mind that a surface must be clean before it can be effectively disinfected. Hose it down with water first, to eliminate mutes and other debris, disinfect it with an appropriate product, and then rinse it thoroughly after about ten minutes (check instruction on disinfectant label). Perches, shelter boxes, and feeding platforms should also be cleaned and disinfected. Outdoor housing is easier to keep clean and disinfected because rain and sunshine help to reduce the number of disease-causing organisms.

- Keep drinking and bathing water clean by checking it daily and thoroughly cleaning the water pans as needed to eliminate the build-up of algae and bacteria.

- Place each water/bathing container in an area that bird mutes won't reach. Don't place it directly under the perch or within reach of a hawk's "projectile" mutes (owls mute downward, while hawks tend to send a stream of mutes outward).

- Establish a feeding platform for your raptor so the food isn't on the ground or in a spot that could be contaminated with mutes. If ants are a concern, one option is to put the feeding platform in a shallow pan of water so the ants can't climb up to get to the food.

4.9b Choosing a Disinfectant

There are several categories of disinfectants, each designed to attack certain types of disease-causing organisms. Each category also has certain properties that might or might not be suitable for the use you have in mind. Read the label carefully before choosing your disinfectant. Here are some considerations to keep in mind:

- Does the disinfectant kill disease agents to which raptors are susceptible? Such agents include strains of Pseudomonas, Salmonella, Streptococcus, Staphylococcus, and Escherichia coli (E. coli). Choose a disinfectant with a wide range of activity.

- Is the disinfectant toxic? Does it have toxic vapors? Is it easily aerosolized? Does it react with other products you use? Is it safe for use around animals and people?

- Is the disinfectant environmentally safe?

- Is the disinfectant surface-friendly or corrosive?

- What is the required contact time for the disinfectant to be effective? Is this time reasonable for your situation?

- What is the best way to apply the disinfectant (spraying, mopping, soaking)? Do you have the necessary tools?

- Does the disinfectant need to be rinsed after application?

- What concentration is used? How long will a supply last? What is its shelf life?

- Is the disinfectant within your budget or too costly?

- What is the optimum ambient temperature and humidity for best performance?

- Is water hardness a concern? The antimicrobial activity of some chemicals decreases when diluted with hard water.

It might seem like a lot to consider, but remember that a disinfectant is a chemical that both you and your bird will come in direct contact with. It needs to be safe and effective.

TRC recommends alternating between two disinfectants, with a dilute bleach solution being one of them. A one to thirty dilution (bleach to water) works well. However, be careful if using bleach. A more concentrated solution can be an irritant both to the birds (with their sensitive respiratory systems) and to people. Other disinfectants commonly used for raptor mews, perches, etc., include Envirocare® Neutral Disinfectant, Roccal D®, Tek-Trol® Disinfectant, and Virkon® S Broad Spectrum Disinfectant.

If you have questions about which disinfectant to use, check with your veterinarian and local hospital to see what they use and recommend for a general disinfectant. Review the Material Data Safety Sheet (MSDS) for the product(s) you choose. If your bird becomes infected with a disease-causing organism, you might need to change products if your general disinfectant is not effective against it. Your veterinarian can help you with this decision.

4.9c Disinfecting Schedule

Once you've chosen your disinfectant, you should create a schedule for disinfecting your raptor's residence areas. Every day, uneaten food and pellets should be removed to prevent the attraction of insects and predators. Then, in non-freezing temperatures, drinking and bathing water should be checked daily and cleaned at least every two to three days. Make sure the ground is free of mutes by hosing down gravel, cement, or artificial turf and raking sand or gravel when needed (raking should only be done when the bird is removed due to dust and fungus that are aerosolized). Sand and gravel should be completely replaced every few years. A thorough disinfecting once per week (depending on the individual species and the number of birds housed together) should suffice. The cleaner you keep your birds' areas and the more often you disinfect them, the less chance there will be of disease.

In snowy climates, small mews can be kept clean by shoveling old, dirty snow out, and replacing it with a thin layer of clean snow.

4.10 COMMON INJURIES ASSOCIATED WITH INADEQUATE HOUSING

You know how your back starts to ache after a long car ride? Or how you puncture your foot by stepping on a nail you forgot to pick up? Or how your lips stick to metal in the bitter cold (don't try this, it does happen!)? Well, think about it. These are all things you can plan for and prevent. You can place a pillow on your car seat, make sure all the nails are picked up, and avoid putting your lips on cold metal. You're aware of these hazards, so you can consciously adjust your behavior to prevent them. The raptors in your care, on the other hand, don't have such control over their living situation; you control everything. As your bird's caretaker, it's your responsibility to eliminate hazards that could cause injury.

In this book, basic guidelines for housing, tethering, and medically caring for permanently disabled raptors are presented. The important thing to keep in mind when reviewing these is that every bird is different in its ability to adapt to captive life. One red-tailed hawk, for example, might do quite nicely free-lofted in a mew lined with mesh, while another red-tail might do better in a doweled enclosure or even tethered. Each individual must be evaluated, and its management plan modified to best meet its needs. A bird's needs can change during its lifetime based on many factors including time of year, change in physical status, age, change in environment, and unknown behavior triggers.

Despite the most carefully thought-out plan, your raptor might injure itself while in your care. The most common injuries that flag a management problem include broken feathers, abraded wrists, bumblefoot, a broken or cracked beak, a bruised cere, a lost talon sheath, and leg abrasions.

4.10a Broken Feathers

Inappropriate housing is the primary cause of damaged (broken or bent) feathers in captive raptors. Your bird's housing or tethering facility must be large enough so that the bird can spread its wings without touching the sides of its enclosure or hitting nearby objects. Never free-loft an excitable bird in a regular mesh enclosure. Tail feathers can break easily if the bird grabs on to the mesh, pushing the feathers through the mesh holes. Such a bird should be tethered or housed in a doweled enclosure.

Remember that a bird's behavior can change with the seasons and many birds increase their activity level during breeding and migratory seasons. This increased activity may require a change in management to prevent a bird from injuring itself or breaking its feathers. For example, a free-lofted bird may need to be tethered during this time for everyone's safety.

Broken feathers can also result from improper placement of perches in the mew. Perches must be

Fig. 4.13
A raptor tethered to a ring perch with jesses that are too long, resulting in broken feathers.
(Gail Buhl)

Fig. 4.14 Wrist wounds.

(a)
Diagram of wrist location. (Gail Buhl)

(b)
Injured wrist in a captive bald eagle.

placed so that the bird can turn around easily and clear the sides of the enclosure with its tail. Beveled perches placed across the mew or angled across a corner can be used if enough turning room is provided. Perches should be high enough so the bird's tail feathers don't touch the ground (many raptors prefer higher perches anyway). Amputees, partial amputees, or ground perchers (burrowing owls, crested caracaras, snowy owls, etc.) should have a ramp or low perch system that gives them access to areas high enough to prevent tail damage.

Another cause of broken feathers is improper tethering. If you tether your bird, make sure that it is away from objects it could hit during a bate, that its perch is safe and won't encourage twisting, and that its jesses are not too long. A raptor must be tethered to a perch that allows it to turn without getting its leash hooked or twisted. Stumps with protruding branches are serious accidents waiting to happen; don't use them for tethering your bird. In addition, if your bird is tethered to a block perch and has exceptionally long jesses, it could wrap its jesses around the perch during a bate and break feathers, or even worse, injure itself (figure 4.13).

4.10b Wrist Wounds

Another common reflection of poor raptor management is wrist wounds (figure 4.14). If a bird is housed in an area that is either too small or constructed out of hard material (cement, solid sheets of wood, chain link), it may injure its wrists when it moves around. This is most likely to occur when the bird is acclimating to a new housing environment or transport carrier, is beginning training, or gets spooked. Injured wrists often become infected, inhibit a bird's ability to fly or comfortably move around, and can lead to permanent joint damage and arthritis. One way to prevent these wounds from occurring in a new bird is to apply wrist protector bandages during the acclimation and training period (chapter 7, Medical Care).

4.10c Bumblefoot

Bumblefoot is a general term referring to injuries to a bird's foot. Most often characterized by abrasions, punctures, corns, or worn spots, bumblefoot usually indicates a housing problem that must be addressed immediately (figure 7.4).

Improper perch size or material, poor placement of perches, cage-clinging tendencies, hard floors (cement, wood), overgrown talons, poor diet, lack of exposure to sun and moisture, and sickness or injuries are all common causes of foot problems in raptors. If your bird develops any form of bumblefoot, closely evaluate its overall health, behavior, and housing, and seek veterinary assistance. Minor foot irritations can turn into major infected wounds in no time. Please refer to chapter 7, Medical Care, for more information on bumblefoot.

4.10d Broken Beak and/or Bruised Cere

Yet another sign that improvements in your raptor management plan are needed is damage to a bird's beak. If your bird's beak is not properly filed (see chapter 6, Maintenance Care), it will grow to a length and thickness that will make it difficult for the bird to tear its food and close its mouth. It is also likely that the tip of the beak will catch on something and break (often quite high, near the face). Therefore, your bird's beak must be filed and reshaped regularly.

Sometimes, if a bird is not housed in a fashion that suits its personality (free-lofted versus tethered, for example), it might hit the sides of the enclosure with its face and break or crack its beak. A damaged beak is often accompanied by a bruised or abraded cere (the colored, fleshy area above the beak). This is a sign that you need to adjust your bird's housing situation.

4.10e Lost Talon Sheath

Have you ever lost a fingernail in a closed door? If you have, you can appreciate how painful it must be to lose the sheath of a talon. The sheath is like a fingernail, protecting a bone (the talon bone) and its blood supply underneath. Every once in a while, a bird can get its talon stuck on a roped perch, in mesh, or on some unknown obstacle, or get frostbite, and lose the sheath from a talon. This usually causes a fair amount of bleeding and is quickly discovered.

As is explained in chapter 7, Medical Care, the talon bone should be cleaned and coated to prevent an infection from occurring. Your veterinarian can help with this. Examine the situation in which the sheath was lost to try to figure out what happened. Perhaps you need to remove something from your bird's mew, change its tether, or adjust its cold-weather housing.

4.10f Leg Abrasions

One final injury common to captive raptors is leg abrasions (figure 4.15). These abrasions are characterized by a loss of feathers (in species with feathering on their lower leg) or scales underneath the anklets, along with abraded or ulcerated skin. Again, several things can result in such an injury.

First, if the anklets applied to your bird's legs are too tight, the

Fig. 4.15
Leg abrasion in a captive raptor.
(Gail Buhl)

legs may swell, causing the anklets to cut into them and create nasty, deep wounds. Second, if the anklets are too narrow, they will not provide adequate support to your bird's legs and will continually abrade them when your bird bates. Third, if the anklets are not the same width on each leg, or the jesses are of different length, uneven pressure and rubbing is applied to each leg when the bird bates and leg injuries easily result. Thus, it is critical to apply anklets of an appropriate width, make sure they are the same width on each leg, and make sure the two jesses are the same length (chapter 5, Equipment). Common barn owls and Northern goshawks tend to be especially susceptible to leg abrasions.

A raptor can also get leg abrasions if its leash becomes twisted when tethered. Birds that end up hog-tied around a perch can develop leg injuries from the jesses rubbing into their skin. Often, twisted birds also have wrist injuries and broken feathers. It cannot be emphasized enough how important it is to frequently check on a tethered bird.

One last thing to mention about leg abrasions is that every now and then, a raptor will shed the scales on its lower leg, revealing a new set of scales underneath. This is normal. However, if leg scales are forced off by the rubbing of anklets, the anklets must be replaced or left off for a while so the leg can heal and scales reform.

4.11 SUMMARY

Suitable housing for your bird must do three basic things:

- Prevent your bird from escaping
- Protect it from predators
- Give it shelter from the elements

Most raptor enclosures, which are called mews, need to be built from scratch. Although raptors can be housed indoors or outdoors, keeping your bird outdoors is usually preferable.

As a raptor manager, you have the responsibility of creating a safe housing arrangement for your bird. Each individual bird has a different tolerance level for captivity. It is your job to continuously evaluate each bird's tolerance and to make sure that its housing arrangement is safe and does not promote injuries. If there is a flaw in the bird's housing, broken feathers, abraded wrists, sore feet, a damaged beak, or leg abrasions may result.

4.12 SUGGESTED READINGS

Buhl, G. and L. Borgia. 2004. *Wildlife in Education: A Guide for the Care and Use of Program Animals.* St. Cloud, MN: The National Wildlife Rehabilitators Association.

Fowler, M. 1993. *Zoo and Wild Animal Medicine.* Philadelphia: B. Saunders.

Henderson, C. 1992. *Woodworking for Wildlife: Homes for Birds and Mammals.* Diane Books Publishing Company.

Jennier, J. 1995. *Enrichment Options – Raptors.* AAZK Forum, 22(10): 424.

Lauback, M. and Rene Lauback. 1998. *The Backyard Birdhouse Book. Building Nestboxes and Creating Natural Habitats.* Pownal, VT: Storey Books.

Martin, S. Enrichment: *What Is It and Why Should You Want It?* Presented at the World Zoo Conference in Pretoria, South Africa, October 1999. www.naturalencounters.com.

McKeever, K. 1987. *Care and Rehabilitation of Injured Owls.* 4th ed. Ontario, Canada: W.F. Rannie.

Russell, A. D., V. S. Yarnych, and A. V. Koulikouskii. 1984. *Guidelines for Disinfection in Animal Husbandry for Prevention and Control of Zoonotic Diseases.* World Health Organization.

Shepherdson, D.J. 1988. *The Application of Behavioural Enrichment in Zoos.* Primate Report 22: 35-42.

Shepherdson, D.J., J.D. Mellen, M. Hutchins. 1998. *Environmental Enrichment for Captive Animals.* Washington DC: Smithsonian Institution.

White, J. D. *Cleaning and Disinfection Processes in Avian Rehabilitation.* Proceedings of the 1993 International Wildlife Rehabilitation Council Conference. IWRA Council.

Table 4.1 Recommended minimum cage sizes for program raptors

Species		Length feet (m)	Width feet (m)	Height feet (m)
American kestrel Boreal owl Burrowing owl Eastern screech owl Merlin Northern saw-whet owl Sharp-shinned hawk Western screech owl	Fully flighted	5 (1.5)	5 (1.5)	7 (2.1)
	Nonflighted or tethered	3 (0.9)	3 (0.9)	3 (0.9)
Cooper's hawk Northern goshawk Northern harrier	Fully flighted or display only	*14 (4.3)	6 (1.8)	7 (2.1)
	Nonflighted or tethered	6 (1.8)	6 (1.8)	7 (2.1)
Broad-winged hawk Common barn owl Long-eared owl Mississippi kite Peregrine falcon Prairie falcon Short-eared owl	Fully flighted	10 (3.0)	8 (2.4)	7 (2.1)
	Nonflighted or tethered	6 (1.8)	6 (1.8)	7 (2.1)
Barred owl Crested caracara Ferruginous hawk Great gray owl Great horned owl Harris's hawk Red-shouldered hawk Red-tailed hawk Rough-legged hawk Snowy owl Swainson's hawk Swallow-tailed kite	Fully flighted	12 (3.6)	8 (2.4)	7 (2.1)
	Nonflighted or tethered	8 (2.4)	8 (2.4)	7 (2.1)
Bald eagle Black vulture Golden eagle Osprey Turkey vulture	Fully flighted (display only)	40 (12.2)	10 (3.0)	9 (2.7)
	Nonflighted, tethered, or fully flighted free-lofted program bird	12 (3.6)	10 (3.0)	9 (2.7)

*It has been TRC's experience that if these species are free-lofted, they are managed better in enclosures that are rectangular (longer and narrower). They need to distance themselves more in order to feel "safe" from the caretaker when he/she goes in to feed them or clean the enclosure.

Table 4.2 Recommended minimum shelter box/hutch sizes for program and display raptors

Species	Depth	Width	Height
Small birds (American kestrel, Eastern screech owl, Northern saw-whet owl, Western screech owl, etc.)	12in (30.5cm)	12in (30.5cm)	12in (30.5cm)
Medium and large birds (barred owl, great horned owl, red-tailed hawk, etc.)	15in (38.1cm)	24in (61.0cm)	24in (61.0cm)
Extra-large birds (Bald eagle, golden eagle, vultures, etc.)	3ft (0.9m)	3ft (0.9m)	3ft (0.9m)

Table 4.3 Recommended perch designs for free-lofted program and display raptors

Species	Surface Size inches (cm)	Cover Type/Size inches (cm)
American kestrel	½ (1.3) dowel ½–1 (1.3–2.5) diameter natural branch 6 (15.2) shelf or block	¼ (0.6) rope Monsanto turf, stadium matting
Bald eagle	Modified 2x4 (5.1x10.2) (figure 4.7a) Natural branches of varying diameters 1.5–3 (3.8–7.6)	3/8 (1) or ½ (1.3) rope
Barred owl	Modified 2x 4 (5.1x10.2cm) (figure 4.7d) Natural branch 1–1.5" (2.5–3.8cm) diameter	3/8 (1) rope
Black vulture	18 (45.7) round diameter block or low square platform 1.5–2 (3.8–5.1) natural branch 2x4 (5.1–10.2) positioned with 4 (10.2) perching surface Large rocks or tree stumps	Monsanto turf or Stadium matting 3/8 (1) rope
Boreal owl	½ (1.3) dowel ½–1 (1.3–2.5) natural branch	¼ (0.6) rope
Broad-winged hawk	1 (2.5) dowel ¾–1 (1.9–2.5) natural branch	¼ (0.6) rope
Burrowing owl	4 (10.2) diameter block perch Natural rocks 4 (10.2), ground stumps, and 4x4 (10.2x10.2) fence posts	Monsanto turf or stadium matting
Common barn owl	Modified 2x4 (5.1x10.2) (figure 4.7c) 1 (2.5) diameter natural branch 6–12 (15.2–25.4) shelf or block	3/8 (1) rope Monsanto turf or stadium matting
Cooper's hawk	1–1.5 (2.5–3.8) dowel 1–1.5(2.5–3.8) natural branch	3/8 (1) rope
Crested caracara	8–12 (20.3–25.4) block Natural stumps, large rocks ¾–1 (1.9–2.5) natural branch	Monsanto turf or stadium matting
Eastern screech owl	1/2 (1.27) dowel ½–3/4 (1.3–1.9) natural branch	¼ (0.6) rope
Ferruginous hawk	Modified 2x4 (5.1x10.2) (figure 4.7b) 1–2 (2.5–5.1) natural branch	3/8 (1) rope
Golden eagle	Modified 2x4 (5.1x10.2) (figure 4.7a) 18 (45.7) block Natural branches 1.5–3 (3.8–7.6) diameter Low Platform 3ftx 3ft (0.9mx0.9m)	3/8 (1) or ½ (1.3) rope Monsanto turf or stadium matting Monsanto turf or stadium matting
Great gray owl	1–1.5" (2.5–3.8cm) dowel 1–1.5" (2.5–3.8cm) natural branch	3/8 (1) rope
Great horned owl	Modified 2x4 (5.1–10.2cm) (figure 4.7b) Natural branches 1–2 (2.5–5.1) diameter	3/8 (1) rope
Harris's hawk	Modified 2x4 (5.1x10.2) (figure 4.7b) 1.25–1.5 (3.2 –3.8) natural branch	3/8 (1) rope
Long-eared owl	¾–1 (1.9–2.5) dowel ¾–1.5 (1.9–3.8) natural branch	¼ (0.6) rope
Merlin	½ (1.3) dowel ½–3/4 (1.3–1.9) natural branch	¼ (0.6) rope

Mississippi kite	3/4–1 (1.9–2.5) dowel ¾–1 (1.9–2.5) natural branch Ground stumps 4x4 (10.2x10.2) fence posts	¼ (0.6) rope
Northern goshawk	¾ (1.9) dowel 1.5 (3.8) natural branch	3/8 (1) rope
Northern harrier	1 (2.5) dowel 1–1.5 (2.5–3.8) natural branch	3/8 (1) rope
Northern saw-whet owl	1/2 (1.3) dowel ½–3/4 (1.3–1.9) natural branch	¼ (0.6) rope
Osprey	Modified 2x4 (5.1cmx10.2cm) (figure 4.7b) 1.5 (3.8) natural branch	3/8 (1) rope Monsanto turf
Peregrine falcon	1ft (0.5m) wide rectangular shelf perch 2x4 (5.1x10.2) positioned with 4 (10.2) perching surface Large rocks	Monsanto turf, stadium matting or coco mat (dry climates) 3/8 (1) rope, Monsanto turf, or stadium matting
Prairie falcon	1ft (0.5m) wide rectangular shelf perch 2x4 (5.1x10.2) positioned with 4" (10.2cm) perching surface Large rocks	Monsanto turf, stadium matting, or coco mat (dry climates) 3/8 (1) rope, Monsanto turf, or stadium matting
Red-shouldered hawk	Modified 2x 4 (5.1x10.2) (figure 4.7b) 1 (2.5) natural branch	3/8 (1) rope
Red-tailed hawk	Modified 2x4 (5.1x10.2) (figure 4.7b) 1.5–2 (3.8–5.1) natural branch	3/8 (1) rope
Rough-legged hawk	Modified 2x4 (5.1x10.2) (figure 4.7c) 1 (2.5) natural branch	3/8 (1) rope
Sharp-shinned hawk	½ (1.3) dowel ½–3/4 (1.3–1.9) natural branch	¼ (0.6) rope
Short-eared owl	¾–1 (1.9–2.5cm) dowel ¾–1.5 (1.9–3.8cm) natural branch 10–12 (25.4–30.5cm) diameter block, or 1ft x1ft (0.3x0.3m) low platform Large rocks, tree stumps, and fence posts	¼ (0.6) rope Monsanto turf or stadium matting
Snowy owl	18x18 (45.7x45.7) low platform. or 12 (30.5) diameter block 1.5–3 (3.8–7.6) natural branch Modified 2x4 (5.1x10.2) (figure 4.7a)	Monsanto turf or stadium matting 3/8 (1) rope
Swainson's hawk	Modified 2x4 (5.1x10.2) (figure 4.7b) 1–1.5 (2.5–3.8) natural branch	3/8 (1) rope
Swallow-tailed kite	3/4–1 (1.9–2.5) dowel ¾–1 (1.9–2.5) natural branch Ground stumps 4x4 (10.2x10.2) fence posts	¼ (0.6) rope
Turkey vulture	18 (45.7) round diameter block low square platform 1.5–2 (3.8–5.1) natural branch Modified 2x4 (5.1–10.2) positioned with 4 (10.2) perching surface Large rocks or tree stumps	Monsanto turf or Stadium matting 3/8 (1) rope
Western screech owl	1/2 (1.27) dowel ½ –3/4 (1.3–1.9) natural branch	¼ (0.6) rope

Table 4.4 Recommended primary perch design for tethered raptors

Species	Perch Type*	Minimum Perch Height (inches/cm)
American kestrel	Bow (¼"/ 0.6cm rope) or block (4-6"/10.2-15.2cm) diameter	6/15.2
Bald eagle	Bow (½"/ 1.3cm rope)	16-18/40.6-45.7
Barred owl	Bow (3/8"/ 1cm rope) or block (6"/15.2cm diameter)	9-11/22.9-27.9
Black vulture	Block (10-12"/25.4-30.5cm diameter)	10-11/25.4-27.9
Boreal owl	Bow (1/4"/ 0.6cm rope) or block (4-6"/10.2-15.2cm diameter)	6/15.2
Broad-winged hawk	Bow (1/4"/ 0.6cm rope)	7-8/17.8-20.3
Burrowing owl	Block (4-6"/10.2-15.2cm diameter)	6/15.2
Common barn owl	Block (6"/15.2cm diameter) or bow (1/4"/ 0.6cm rope)	6/15.2
Cooper's hawk	Bow (3/8"/1cm rope)	10-11/25.4-27.9
Crested caracara	Bow (1/2"/ 1.3cm rope) or block (8-10"/20.3-25.4cm diameter)	10-11/25.4-27.9
Eastern screech owl	Bow (1/4" /0.6cm rope) or block (4-6"/10.2-15.2cm diameter)	6/15.2
Ferruginous hawk	Bow (3/8"/ 1cm rope) or block (8-10"/20.3-25.4cm diameter)	10-11/25.4-27.9
Golden eagle	Block (12"/30.5cm diameter)	16-18/40.6-45.7
Great gray owl	Bow (3/8"/ 1cm rope) or block (8-10"/20.3-25.4cm diameter)	10-11/25.4-27.9
Great horned owl	Bow (3/8"/ 1cm rope)	10-11/25.4-27.9
Harris's hawk	Bow (3/8"/ 1cm rope)	10-11/25.4-27.9
Long-eared owl	Bow (1/4"/0.6cm or 3/8"/1cm rope)	6/15.2
Merlin	Bow (1/4"/ 0.6cm rope) or block (4-6"/10.2-15.2cm diameter)	6/15.2
Mississippi kite	Block (6"/15.2cm diameter) or bow (3/8"/ 1cm rope)	7-8/17.8-20.3
Northern goshawk	Bow (3/8" / 1cm rope)	10-12/25.4-30.5
Northern harrier	Block (6"/15.2cm diameter) or bow (3/8"/ 1cm rope)	11/27.9
Northern saw-whet owl	Bow (1/4"/ 0.6cm rope) or block (4-6"/10.2-15.2cm) diameter	6/15.2
Osprey	Bow (3/8"/ 1cm rope) or block (8-10"/20.3-25.4cm) diameter	10-11/25.4-27.9
Peregrine falcon	Block (5-7"/12.7-17.8cm diameter)	9-10/22.9-25.4
Prairie falcon	Block (5-7"/12.7-17.8cm diameter)	9-10/22.9-25.4
Red-shouldered hawk	Bow (3/8"/ 1cm rope)	10-11/25.4-27.9
Red-tailed hawk	Bow (3/8"/ 1cm rope)	10-11/25.4-27.9
Rough-legged hawk	Bow (3/8"/ 1cm rope) or block (7"/17.8cm diameter)	10-11/25.4-27.9
Sharp-shinned hawk	Bow (1/4"/0.6cm rope) or block (4-6"/10.2-15.2cm diameter	6/15.2

Short-eared owl	Bow (1/4"/0.6cm or 3/8"/1cm rope) Block (6"/15.2cm)	6/15.2
Snowy owl	Bow (3/8"/1cm rope) or Block (8-10"/20.3-25.4cm diameter)	10-12/25.4-30.5
Swainson's hawk	Bow (3/8"/1cm rope)	10-11/25.4-27.9
Swallow-tailed kite	Block (6"/15.2cm)	14-15"/38.1
Turkey vulture	Block (10"/25.4cm diameter)	10-12/25.4-30.5
Western screech owl	Bow (1/4"/0.6cm rope) or block (4-6"/10.2-15.2cm diameter)	6/15.2

* The bow perches listed here are standard 5/8"(1.6 cm) steel bow perches available from falconry suppliers listed in Appendix D. Larger diameter steel perches or rod iron perches can be custom made, but the size of the perch wrapping would have to be adjusted to prevent the finished diameter from getting too large for the size of the bird's feet.

Table 4.5 Raptor species that can be housed safely together for display or program use.

Species	American kestrel	Bald eagle	Barred owl	Black Vulture	Boreal owl	Broad-winged hawk	Burrowing owl	Common barn owl	Cooper's hawk	Crested caracara	Eastern screech owl	Ferruginous hawk	Golden eagle	Great gray owl	Great horned owl	Harris's hawk	Long-eared owl	Merlin	Mississippi kite	Northern goshawk	Northern harrier	Northern saw-whet owl	Osprey	Peregrine falcon	Prairie falcon	Red-shouldered hawk	Red-tailed hawk	Rough-legged hawk	Sharp-shinned hawk	Short-eared owl	Snowy owl	Swainson's hawk	Swallow-tailed kite	Turkey vulture	Western screech owl
American kestrel	X																																		
Bald eagle		X																																	
Barred owl			X																																
Black vulture				X						X																								X	
Boreal owl					X																														
Broad-winged hawk						X																								X					
Burrowing owl							X																												
Common barn owl								X																											
Cooper's hawk									S																										
Crested caracara				X						X																X								X	
Eastern screech owl											X																								X
Ferruginous hawk												X																							
Golden eagle													X																						
Great gray owl														X																					
Great horned owl															X																				
Harris's hawk																X																			
Long-eared owl																	X													X					
Merlin																		S																	
Mississippi kite																			X														X		
Northern goshawk																				S															
Northern harrier																					X														
Northern saw-whet owl																						X													
Osprey						X																	X			X	X	X				X			
Peregrine falcon																								X											

Species	American kestrel	Bald eagle	Barred owl	Black Vulture	Boreal owl	Broad-winged hawk	Burrowing owl	Common barn owl	Cooper's hawk	Crested caracara	Eastern screech owl	Ferruginous hawk	Golden eagle	Great gray owl	Great horned owl	Harris's hawk	Long-eared owl	Merlin	Mississippi kite	Northern goshawk	Northern harrier	Northern saw-whet owl	Osprey	Peregrine falcon	Prairie falcon	Red-shouldered hawk	Red-tailed hawk	Rough-legged hawk	Sharp-shinned hawk	Short-eared owl	Snowy owl	Swainson's hawk	Swallow-tailed kite	Turkey vulture	Western screech owl
Prairie falcon																								X											
Red-shouldered hawk																										X		X						X	
Red-tailed hawk									X														X				X	X				X		X	
Rough-legged hawk																										X	X	X				X		X	
Sharp-shinned hawk																													S						
Short-eared owl																	X													X					
Snowy owl																															X				
Swainson's hawk																											X	X				X		X	
Swallow-tailed kite																			X														X		
Turkey vulture			X		X				X																		X	X				X		X	
Western screech owl											X																								X

s = same sex

Chapter 5: EQUIPMENT

When you are tethering or training your raptor, you'll need to have special equipment as well as handling and training tools. Your bird will need a pair of anklets and jesses, a swivel, and a leash, and you'll need a glove and/or a gauntlet. In addition, you must have a scale to weigh your bird, and a spray bottle. Optional pieces of equipment include a hood, a whistle or clicker, a lure, and a transmitter. All of this equipment can be purchased from a variety of individual or commercial suppliers (appendix D). If you're handy, however, and have access to the needed supplies, you can easily make some pieces of equipment, such as jesses and leashes.

A bird's equipment is a very important component of its overall management. For program birds, the anklets, jesses, leash, and swivel function to allow your educator to be comfortably restrained on the glove, safely tethered to a perch for housing or display, and to prevent accidental escape. These pieces of equipment must be monitored carefully for wear, as old, faulty equipment can lead to serious injury and in extreme situations, death. A scale is a necessity for any bird you have in your care. Display birds and program birds alike need to be weighed regularly to determine their overall health and well being (chapter 6, Maintenance Care).

5.1 TYPES OF EQUIPMENT

There are several pieces of equipment you may need for your raptor. Please check with your individual state for any regulations that might restrict the type of equipment you use.

5.1a Jesses and Anklets

Jesses are straps that are fastened around a raptor's lower legs, above its feet. They allow a handler to secure a bird on a gloved hand and are essential for tethering a raptor to a perch (figure 5.1). There are two types of jesses, aylmeri and traditional.

Fig. 5.1
Aylmeri jesses, a swivel, and leash properly applied to a red-tailed hawk. (Ron Winch)

Aylmeri jesses

Aylmeri jesses consist of two individual parts, an anklet that wraps around each lower leg, and a strap (the actual jess) which slips through a grommet in each anklet (figure 5.1). This system allows the jesses to rotate in the grommets, decreasing the likelihood that a bird will become twisted or "hog-tied" if tethered. Aylmeri jess straps are usually made out of a single layer of leather, but braided leather, nylon cord, and braided nylon straps are also used.

Aylmeri jesses, in contrast to the traditional type, are a little more "wearer-friendly." A bird that escapes can remove the jess straps merely by pulling them through the grommet, preventing them from catching on objects and causing injury. In this case, only the anklets remain, and they rarely cause problems.

Traditional jesses

Traditional jesses consist of a single piece of leather that forms an anklet around the bird's leg and leaves a length of strap for attaching a swivel and leash (figure 5.2). As you will see, traditional jesses are easier to make, attach, and remove (if a bird's leg becomes irritated), but once you put them on, the bird cannot pull them off. This is a critical factor if your bird escapes, as the straps can catch on objects, trapping your bird and resulting in injury or death. Also, a tethered bird wearing traditional jesses is more likely to get twisted, since the jess straps cannot rotate with the bird's movements. TRC does not recommend their use for permanently disabled raptors, and some states prohibit them. Traditional jesses are also not for use during free-flight demonstrations.

Whichever type of jess system you decide on, keep in mind that the anklet around the leg must be tight enough that it won't slip down over your bird's foot or up over the hock (ankle) joint, but not so tight that it can't freely rotate around your bird's leg. Also, make sure you always have an extra pair of jesses and anklets handy in case one or both need to be replaced. Never use only one jess, or handle/tether a bird with weak or faulty jesses. Leather jesses can get dry and crack, stretch, become thin under the button and at the bottom slit. Nylon jesses can fray.

Instructions on how to make and apply jesses are provided later in this chapter.

5.1b Swivel

A swivel, which connects a bird's jesses to its leash (figure 5.1), consists of two stainless-steel rings separated by a solid piece of steel, or for small raptors, one stainless steel ring and one rounded clip. Avoid triangular clips as these tend to open more easily on their own

Fig. 5.2
Conventional traditional jesses applied to the legs of an owl. (Gail Buhl)

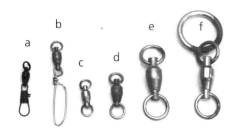

Fig. 5.3
Swivels for small (a-c), medium (d-e), large (e), and extra-large raptors (f). (Ron Winch)

or break at the bend (figure 5.3). One ring (the swivel end) rotates, and the other end is stationary. Fishing swivels are most commonly used and can be purchased at sporting goods stores. They are made in various tensile strengths. TRC recommends 300-pound Sampo® swivels for small and medium-sized birds, 500-pound for large birds, and 800- or 1,200-pound for extra large birds (eagles).

For small and medium-sized birds, a double swivel system is sometimes used to prevent the jesses from twisting (figure 5.4). One variation of this system consists of two tiny clip swivels, one of which clips to each jess. The rings of each swivel then attach to the ring of a third, larger swivel (like keys on a key ring). The leash is then passed through the free ring of this third swivel. A second variation eliminates the third swivel and just has a leash pass directly through the ring on each clip swivel. The double swivel system has a lot of hardware for a small bird but can be effective if twisting is a major problem.

Your bird should wear a swivel if it is being held on the fist or tethered. Never leave a swivel on a bird that is free-lofted. The swivel effectively attaches the bird's legs together, limiting its ability to move. It can also get hooked on objects, tangling your bird and causing injuries.

Attaching a swivel to jesses

To attach a two-ring swivel, pass the two tapered jess ends through the "stationary" ring of the swivel (the ring that does not rotate), past the point of the slits (figure 5.5a). Then, pass the rotating ring of the swivel through the two slits and pull it snugly (figure 5.5b). Clip swivels should only be used with jesses that are made with a single hole at the tapered end (recommended for small raptors). To attach a clip swivel, put the clip through both holes at the tapered jess ends and fasten.

Even though swivels are constructed out of solid materials, they can also show wear with time and their condition must be monitored (figure 5.6). The rotating ring can stop turning or become too loose, the solder points on the rings can get thin and potentially break, and the clips on clip swivels can become faulty. Also, some individuals may repeatedly bite at their swivels and develop sores in the corners of their mouths or chip their beaks. If this happens, you can create a protective "sock" to slip over the swivel (figure 5.7).

5.1c Leash

Whenever a bird is taken out of its mew, held on the glove for a program, or tethered to a perch, you must use an appropriate leash to prevent it from escaping. Leashes are typically made out of a sin-

Fig. 5.4
Double swivel system for tethering small raptors. (Ron Winch)

Fig. 5.5
Attachment of a swivel to the jesses. (Ron Winch)

(a)
Jess ends passed through the stationary ring of a swivel past the point of the slits.

(b)
Rotating end of a swivel passed through jess slits and pulled snugly.

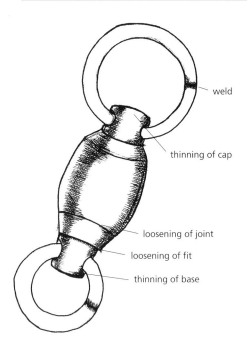

Fig. 5.6
Wear spots on a swivel. (Jen Veith)

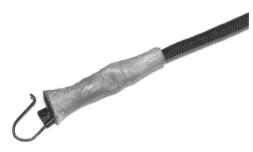

Fig. 5.7
Protective sock placed over swivel.
(Ron Winch)

gle strap of leather, braided leather, nylon, or braided nylon. Leather leashes are made with a button on one end that is large enough so it won't pass through the swivel. Nylon and braided nylon leashes are made either with a button on one end or a slit for attaching to the swivel (figure 5.8). Both types of leashes should be applied by passing them through the free-rotating end of the swivel once it is attached to the jesses (figure 5.1).

Like the other pieces of equipment mentioned so far, leashes can also wear with time and must be monitored carefully. Leather leashes tend to get dry and crack, especially under the button, and have a greater tendency to break than do nylon leashes. Leather leashes should be treated regularly with a leather conditioner.

5.1d Glove/Gauntlet

Anyone handling a raptor, in any situation, should always wear a glove. First, a raptor's sharp beak and talons — even those of the "best-trained" raptor — can make quite a nice puncture in your hand or arm. Second, a glove provides a nice, comfortable perching surface and teaches your bird that when you're wearing the glove, the bird will be handled positively. Third, wearing a glove shows the public that you have respect for your bird, no matter what its size. It promotes the awareness that raptors are trained ambassadors, not pets.

Raptor handlers use several different types of gloves, including welder's gloves, the leather work or garden gloves available at most hardware or liquidation stores, and custom-made leather gloves produced specifically for working with birds or mammals (appendix D).

Gloves come in different sizes, thicknesses, and lengths; you should choose the right combination based on the size of your bird and the size of your hand. For small raptors, the most appropriate gloves extend to just above your wrist; for medium and large birds, to just below your elbow; and for extra large birds, to past your elbow (gauntlets are available that fit over most of your arm).

Some custom gloves are made with an optional small ring or ring/clip combination near the wrist. These features are handy for attaching a leash, acting as an extra safeguard against having your bird slip through your fingers and fly away — leash, swivel, and all. Many people who work with birds regularly have their gloves custom made (appendix D). TRC recommends making a small investment to purchase one, as it will provide extra comfort for both you and your bird.

Just like the other pieces of leather equipment, your glove should be kept clean and conditioned. If you feed your bird on the fist, make sure to wash your glove often and apply a leather conditioner. Gloves that are not cleaned become stiff and a breeding site for bacteria. It is also a good idea to have a separate glove for handling during programs. A clean glove, without dried blood and animal fur, presents better to the public. Lastly, it is strongly recommended to use a different type and color of glove for restraining your bird during maintenance and medical exams (chapter 6, Maintenance Care). You always want your bird to feel "safe" while perched on your program glove.

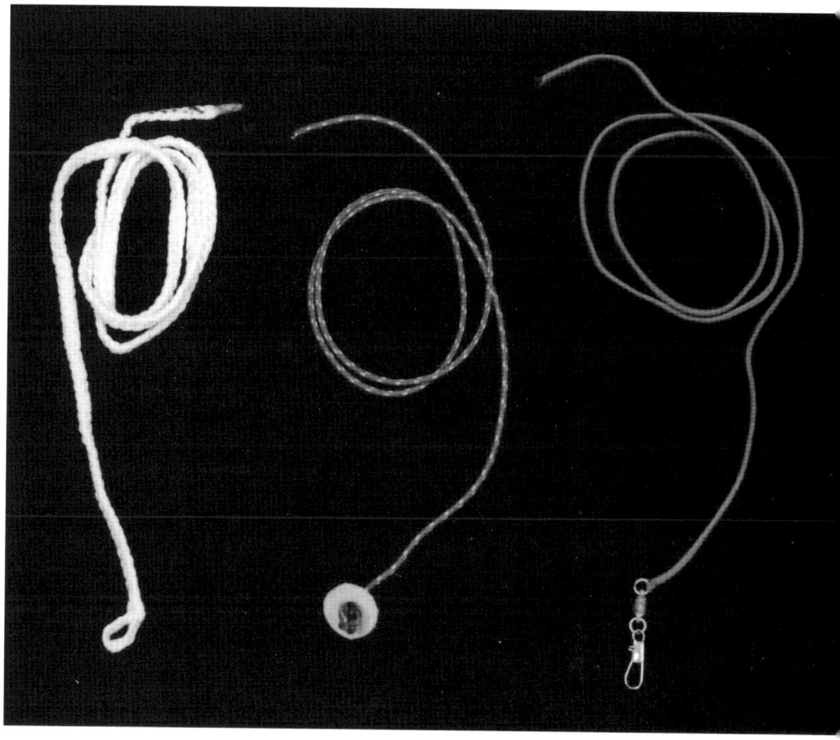

5.1e Scale

TRC cannot overemphasize the importance of owning a scale if a raptor is under your care. Whether or not you'll be flying your bird for audiences, a scale is a very important tool for keeping track of your bird's overall health. More information on this is provided in chapter 6, Maintenance Care.

Scales can get pretty pricey, but all you need is a basic model with some type of perch, which you can make easily. Digital scales are especially nice, offering you weights in one-, two-, or five-gram (28.3, 56.7, or 141.7 oz) increments. See appendix D for recommended suppliers.

5.1f Spray Bottle

Any clean, plastic spray bottle that hasn't previously held any chemicals will suffice. Keep it handy for misting your bird if it becomes hot during a handling session, when tethered for display, or is warm during transport. Direct a light mist to exposed skin surfaces, such as the legs and feet or the head and mouth. Don't direct a coarse spray into your bird's mouth or onto any part of your bird's body.

5.1g Hood

A raptor hood is a custom-designed cap that fits over the head and eyes of a raptor to keep it calm (figure 5.9). It comes in two basic designs, Indian and Dutch. An Indian hood is light and flexible; a Dutch

Fig. 5.8
Leashes for holding or tethering raptors (button and slit). (Ron Winch)

Fig. 5.9
A typical hood that can be placed over the head of an excitable hawk to keep it calm. (Gail Buhl)

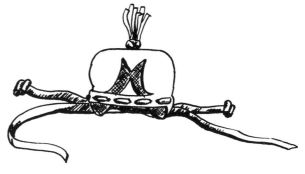

hood is slightly heavier and stiffer. Hoods are most often used for larger falcons and accipiters, but hoods for buteos and eagles are also available. Hoods are neither available nor necessary for owls or vultures.

A hood should only be used for relatively short periods of time, such as the time required to transport a bird within on-site locations or to off-site locations, not for 24 hour-a-day behavior control. Also, a hood must fit your bird's head properly to prevent injuries. The most common injuries include severe eye damage from a hood that rubs on a bird's eyes, and abrasions on a bird's cere, nares, and lips from a bird continually scratching at an uncomfortable hood. If you are interested in training and working your bird with a hood, please contact a local falconer or experienced educator for assistance. Your state or federal wildlife office (appendix A) will have a list of falconers and licensed educators in your area that can help you.

5.1h Whistle or Clicker

A whistle or clicker is often used during training. More information on this is provided in chapter 8, Training.

5.2 MAKING AND ATTACHING EQUIPMENT

5.2a Supplies

As mentioned previously, anklets, jesses, and leashes are easy to make out of leather or nylon rope. All you need is a little time and a few basic supplies (figure 5.10). Vendors for the supplies listed can be found in appendix D.

Leather
There are numerous types of leather that can be used to make high-quality equipment (anklets, jesses, leashes) for your bird. The material you choose should be soft and pliable but not stretchy, and strong enough that it does not rip easily (some parts of a hide are stronger than others). Leather is usually slightly stretchy going with the grain and firm going against the grain. When making equipment cut the leather lengthwise against the grain so anklets, jesses and leashes won't stretch. For smaller birds, we recommend leather from 1/8 inch down to 1/16 inch thick. Kangaroo leather works well for small and medium-sized birds, and thicker cowhides are good for large and extra large raptors. Leather can be purchased from leather shops, shoe-repair shops, or falconry-equipment suppliers.

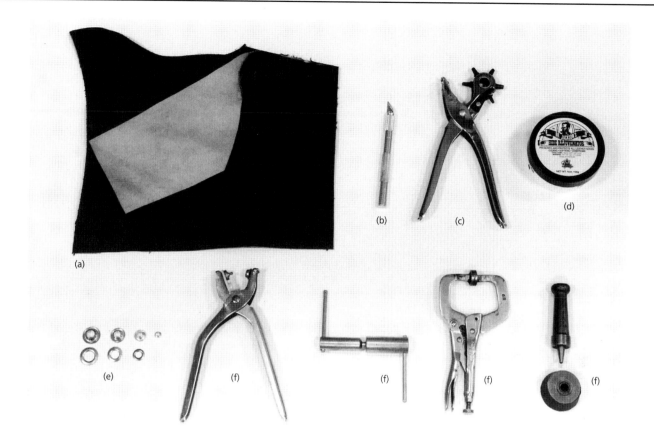

Leather conditioner

It seems as if there are almost as many types of leather conditioners as there is leather. Anytime you apply leather equipment to your bird, it should be conditioned regularly to keep it from drying and cracking. TRC uses Dr. Jackson's Hide Rejuvenator (Tandy Leather) and Jess Grease (by D.B. Scientific). Other types of leather conditioners can be purchased from your leather supplier.

One note of caution: From TRC's experience, pure neatsfoot oil does not protect kangaroo leather and even appears to weaken it.

When you first make anklets, jesses and/or leashes, the final step should be to condition the leather. When applying conditioner to your new equipment, you can massage it into the leather with your hands or a cloth and then heat it in the microwave on high for five to ten seconds. This helps the leather to absorb the conditioner, making it softer. Once the equipment has been put on your bird, conditioning the leather should be continued on a regular basis (biweekly or monthly). This can be done with the bird on your gloved hand. With your free hand, rub conditioner on leather anklets and jesses, making sure to slightly lift the jesses to apply conditioner under the button. Do not put conditioner on the inside of anklets worn by birds with feathered legs. The conditioner will stick to the feathers and potentially cause the anklet to have reduced

Fig. 5.10
Equipment needed for making jesses or a leash for an education raptor:

(a) leather
(b) blade or cutting tool
(c) leather punch
(d) leather conditioner
(e) grommets
(f) grommet pliers.

(Ron Winch)

Fig. 5.11
The two-part grommet system.
(Jen Vieth)

(a)
Grommet hat.

(b)
Grommet ring.

mobility around the leg. Also, leg feathers that are caked with substances can result in skin irritation and abrasions.

Leather punch

If you are making jesses out of leather, you'll need a punch to create holes in the leather. Leather punches can be purchased from leather-supply stores or some fabric shops.

Nylon

Some people prefer to use jess straps made out of nylon rope. Parachute cord has proven to be a good nylon strap material. It's available from sporting-goods and outdoor stores and comes in a variety of colors and diameters.

Grommets

If you're going to make aylmeri anklets and jesses, you'll need grommets for attaching the anklets around your bird's legs. They consist of two independent parts, a hat and a ring that are secured together with a grommet setter (figure 5.11). Grommets come in different sizes, and which size you need is determined by the width of the anklet you're making. Recommended grommet sizes are listed in table 5.1. Eyelets used for small birds are sold in fabric stores. Grommets are available from leather stores, hardware stores, and other suppliers listed in appendix D.

Grommet-setter

If you're making aylmeri anklets, you'll also need a setter for attaching the grommets. These setters are specific to different-sized grommets, so you might need more than one if you have birds that require different grommets. TRC recommends purchasing grommet pliers for ease of applying anklets. Twist setters, and hammer/anvil-setters are also available but are more awkward to use. Grommet-setters can be purchased from your grommet supplier and/or falconry supply companies.

5.2b Instructions

To apply equipment to a new untrained bird, you will need to "cast" it — grab the bird and restrain it on its back (chapter 6, Maintenance Care). Once your avian educator has been trained, you might be able to fasten new anklets with the bird positioned on a perch or a handler's glove. This is most easily done with birds that are trained to a hood.

Aylmeri jess system

The aylmeri jess system consists of an anklet, which is wrapped around a bird's lower leg, and a strap (jess) that passes through a hole in the anklet. There are two kinds of aylmeri jess systems used.

The most common type is the conventional aylmeri jess system. This equipment consists of an anklet that is securely fastened around the bird's leg by a grommet.

The modified aylmeri jess system consists of an anklet that is not "permanently" fastened around the leg. The jess straps pass through a grommet on each end of the anklet and hold the anklet in place. These anklets can be more easily removed and are reusable.

Recommended sizes of anklets and jesses are listed in table 5.1. A length for anklets is not provided because every bird has a different diameter leg, and the anklet must be fitted so it is neither too loose nor too snug.

Anklets:

Make conventional aylmeri anklets: (figure 5.12)

1. Cut a piece of leather according to the measurements given in table 5.1. The piece should be long enough to fit comfortably around your bird's leg and allow room for attaching the grommet. To be safe, cut it a little long; you can always trim it later.

2. Using a leather punch, create a hole at one end. The hole must be large enough so that the size grommet you choose (see table 5.1 for recommended sizes) will fit into it snugly (figure 5.12a) and be positioned so that leather surrounds the grommet's diameter.

3. Take the grommet hat and fit it through the hole (figure 5.12b).

4. Round off the leather on the end with the grommet.

5. Wrap the leather around your bird's leg and "fit" the anklet. Size it so that it can't slip over your bird's foot or hock (ankle) joint but is loose enough to rotate freely around your bird's leg. Press the grommet hat into the leather to make a mark where you need to punch a second hole, on the other end of the leather. (figure 5.12c).

6. Punch a hole where you made the mark and round off this end with scissors.

7. Between the two holes you made, make small cuts on the top and bottom edges of the leather. This adds flexibility to the

Fig. 5.12
Making and applying conventional aylmeri anklets. (Gail Buhl)

(a)
Diagram of an anklet showing rounded ends, holes for grommet placement, and small slits for flexibility.

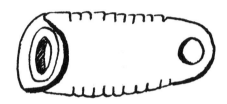

(b)
Grommet placed in one hole.

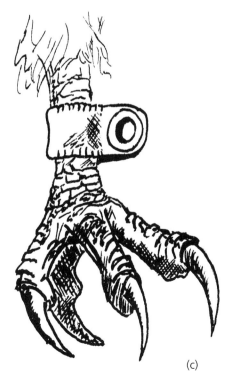

(c)
Anklet properly fitted around a raptor's leg.

Fig. 5.13
Making and applying modified aylmeri anklets. (Jen Veith)

(a)
Inserting the first grommet and making a middle T-slit in the anklet.

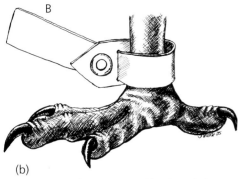

(b)
Measuring the correct anklet length by wrapping it around the lower leg, inserting end B through the slit, and lining up both grommet holes.

(c)
Making the second grommet hole, and applying the grommet, trimming the end, adding notches and slits.

leather so it doesn't cause injury when the bird moves its leg. However, be careful not to make slits too close to the grommet.

8.   Massage conditioner into the leather to soften it.

Apply conventional aylmeri anklets: (figure 5.12c)

1.   Wrap the anklet around your bird's leg. It does not matter which side of the leather (smooth or rough) is closest to your bird's skin, this is personal preference. However, be consistent on both legs.

2.   Place a grommet hat through the two holes you have made. TRC prefers to place the grommets so the smooth side (the back of the grommet hat) faces behind the bird when the grommet is between the bird's legs. This determines how the jesses will be positioned and is just personal preference. However, make sure to fasten the grommets with the smooth side (back of the hat) facing the same direction on both legs.

3.   Place the ring section of the grommet over the grommet hat.

4.   Crimp the grommet, using the setter, and make sure it has been properly attached. If you aren't familiar with the use of this tool, practice before you try to apply the anklets. It is a good idea to have a second measured pair of anklets ready in case you have problems crimping the grommet. Sometimes grommet setters don't function properly (they crimp the grommet unevenly or even crack it) and the resulting damage to the anklets prevents them from being used. Also, if applying anklets to feathered legs, it is a good idea to lightly wet the leg feathers to prevent them from getting caught in the grommet during crimping.

5.   Make sure that the anklet won't slip over the foot or hock (ankle) joint, but that it can rotate around the leg without resistance.

Make modified aylmeri anklets: (figure 5.13)

1.   Cut a long piece of leather at the width recommended in table 5.1.

2.   Punch a hole at one end ("A") to fit the desired grommet (figure 5.13a).

3. Place the hat of the grommet through the rough side of the leather and the ring of the grommet on the smooth side of the leather. Crimp the grommet closed using your grommet setter (figure 5.13a).

4. Make a "T" slit by first making a vertical slit in the leather positioned so it is approximately the distance of one grommet diameter away from the grommet hole just made (figure 5.13a). To make this slit, punch two small holes using the smallest setting on your leather punch and cut a slit between them using a sharp blade. The holes help prevent the leather from tearing above and below the slit. Then in the middle length of the vertical slit, make a horizontal slit moving away from hole A (5.13a).

5. Wrap the anklet around your bird's leg, take end B, pass it through the slit and pull it until the anklet reaches the desired fit around the leg. It should be loose enough to freely rotate around the leg but not too loose that it will slip over the foot or above the hock joint. At this point, end B may be longer than end A (figure 5.13b).

6. Without moving ends A or B, press the grommet in end A against the leather in end B and make a mark on end B where you need to punch a hole for the second grommet. The two grommets must line up well (figure 5.13b).

7. Remove the anklet from the bird. Using your leather punch, make a hole in the leather where you made the mark. Place the grommet hat through the rough side of the leather and the grommet ring on the smooth side of the leather. This will ensure that the jess will pass through the smooth side of the grommet when on your bird. Secure the grommet with your setter (figure 5.13c).

8. After passing end B through the slit, line up the grommets so they are flush and trim end B to line up with end A.

9. With both grommets lined up, make two small notches on end B where it rests in the slit. These will help make B fit comfortably into A when through the slit (figure 5.13c).

10. Between the notches and slit, make small cuts on the top and bottom edges of the leather (figure 5.13c). This adds flexibility to the leather so it doesn't cause injury when the

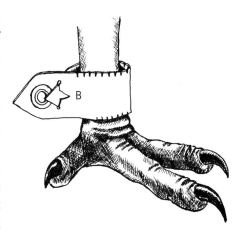

(d)
Modified aylmeri anklets properly applied to a raptor.

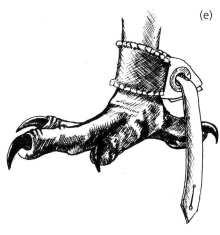

(e)

Fig. 5.14
Making leather aylmeri
jesses. (Gail Buhl)

(a)
Cut leather to desired
length, taper one end 2,
make two folds on oppo-
site ends, and punch
through each fold.

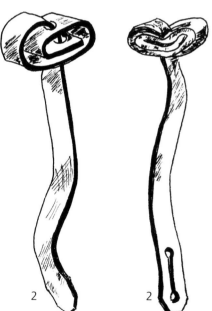

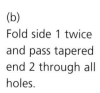

(b)
Fold side 1 twice
and pass tapered
end 2 through all
holes.

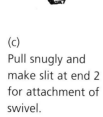

(c)
Pull snugly and
make slit at end 2
for attachment of
swivel.

bird moves its leg. However, be careful not to make slits too close to the grommet.

11. Massage conditioner into the leather to soften it.

Apply modified aylmeri anklets:

1. Place the anklet around your bird's leg with the smooth side of the leather facing out. Take end B and put it through the slit (figure 5.13d). Pull end B until the grommets line up and end B slips into the notches.

2. Take your jess and pass it through the smooth side of the grommet to secure the anklet (figure 5.13e).

Jesses:
Make leather aylmeri jesses: (figure 5.14)

1. Take the desired length and width of leather and taper one end to a point, using a blade or other sharp instrument. Both jesses of a pair must be the same length.

2. Make two to three folds (in a roll fashion) on the non-tapered end. The folds should be about ¼ inch (0.6 cm) long for small birds, ½ inch (1.3 cm) long for medium and large extra-birds, and 1 inch (2.5 cm) long for large birds. These folds will create the button, which must be larger than the grommet you use to attach the anklet.

3. Using a leather punch, punch a hole through the center of each fold. The hole must be big enough to allow you to pass the tapered end of the jess through, but not so large that the jess tears at its width in the button (figure 5.14a).

4. With the leather folded like a button, pass the tapered end through all the holes (beginning at the top of the button) and pull snugly (figure 5.14b).

5. At the tapered end, punch a small hole at the point where the taper ends and the widest part of the leather begins. Make a second hole towards the button at the following rec-ommended distance away from the first hole: ½ inch (1.3 cm) for small and medium-sized birds, 1 inch (2.5 cm) for large birds, and 1½ inches (3.8 cm) for extra large birds. Then take a blade or scissors and make a slit between the

two holes. This slit needs to be large enough for a swivel to pass through (figure 5.14c). It is important that for a pair of jesses, the position of the slit in each jess is the same. Soften the jess by massaging it with a leather conditioner.

6. Slip the jess through the smooth side of the grommet.

7. Repeat steps 1–6 to make the second jess. Make sure that both jesses in a pair are the same length. If they are not, they will cause uneven pressure points on your bird's legs if it bates and can result in severe leg injuries.

Make nylon cord aylmeri jesses: (figure 5.15)

1. Take a 2.5–3 foot (0.76-0.91 m) piece of nylon cord and make a small horizontal slit halfway through the diameter of the cord, approximately 1½ inch (3.8 cm) from one end.

2. Remove the inner fibers that are now exposed (figure 5.15a).

3. Melt the end through which you pulled the inner fibers, then feed this end of the nylon through a second slit made at the middle length of the cord (a coat hanger or hemostat might help with this). Make sure to leave a small loop for the swivel to pass through at the end (figure 5.15b-d). Note: Melt nylon in a well-ventilated area. Melted nylon is hot and can cause burns.

4. Trim the end you pulled through until the jess is about 1–1.5 inch (2.5–3.8 cm) longer than the desired final jess length (table 5.1). On this end, make a knot and burn the tip to melt it (figure 5.15e–g). The knot must be large enough so that it won't slip through the grommet. A washer, circular piece of leather, or piece of shrink tubing can be applied under the knot for extra protection against slipping or wear on the knot.

5. Repeat steps 1–4 to make the companion jess for the other leg. Make sure to make both jesses of a pair the same length. Uneven jesses can lead to severe leg injuries if a bird bates.

If you decide to use aylmeri jesses and your bird is free-lofted, it may repeatedly pull the jesses out while it is getting used to them. To make the jesses harder for your bird to remove, there are several jess modifications you can use on the side of the anklet opposite the button. These include making "wings" under the buttons of

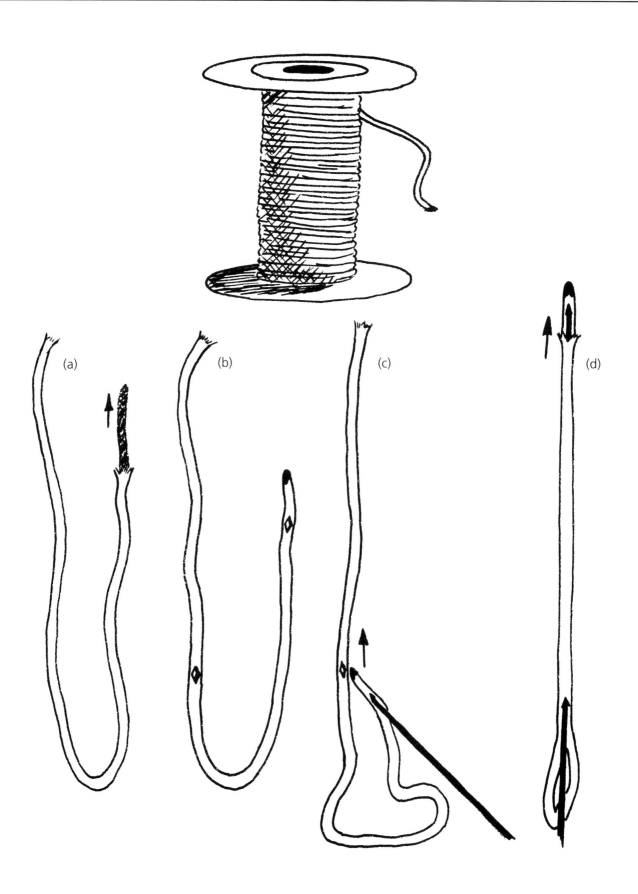

(a)

(b)

(c)

(d)

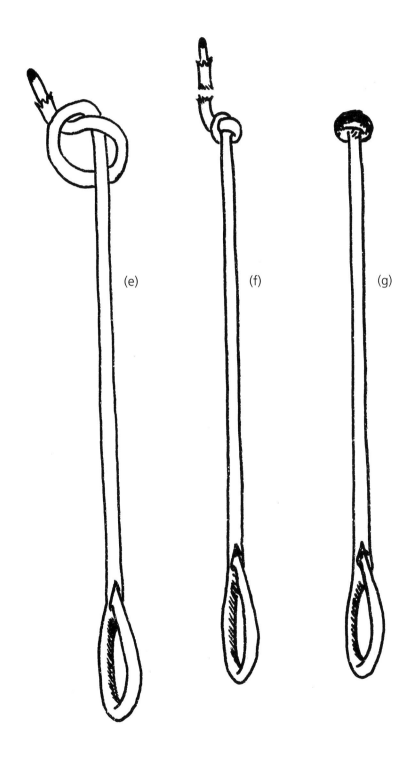

(e)   (f)   (g)

Fig. 5.15
Making nylon aylmeri jesses.
(Gail Buhl)

(a)
Inner fibers removed from cord.

(b,c,d)
The end burned and passed
through cord with piece of stiff
wire until a small loop remains
(large enough to pass the ring
of a swivel).

(e,f,g)
Knot tied on free end and end
burned to prevent unraveling.

Fig. 5.16
Jess modifications to prevent a raptor
from removing aylmeri jesses.
(Jen Veith)

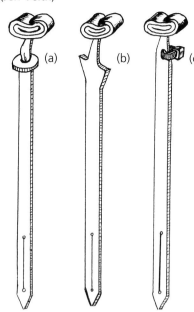

(a) Leather button.
(b) Wings under knot.
(c) Plastic cable tie.

Fig. 5.17
A conventional traditional jess and
its application. (Gail Buhl)

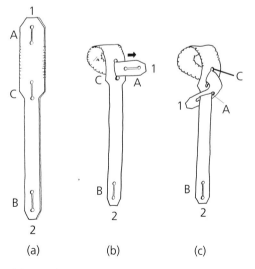

(a) Jess design.
(b) Wrap end 1 around a bird's lower leg and
pass it through slit C. This will expose slit A.
(c) Pass end 2 through slit A.

leather jesses, or applying a flat round piece of leather snuggly to each jess (figure 5.16) to create resistance. If nylon jesses are used, some people tie an extra knot or secure a cable tie to the jesses on the other side of the anklets. However, if your raptor is used for free-flight demonstrations, do not use any jess modifications. If the bird takes off, it needs to be able to pull out its jesses to decrease the possibility of getting snagged or hung up.

Traditional jesses

The traditional jess system consists of a single piece of leather that is made to circle around a bird's lower leg and provides a strap that hangs down for attachment of a leash and swivel (figure 5.2). Recommended widths are those listed for anklet widths in table 5.1. There are two designs for traditional jesses: the conventional design (figure 5.17) and a modified design that keeps the anklet size stable created by John Karger, Last Chance Forever/The Bird of Prey Conservancy (figure 5.18).

Make conventional traditional jesses: (figure 5.17a)

1. Cut the desired width and length of leather, and taper it at both ends.

2. Using a leather punch, make two pairs of holes (A, B), one at each end of the leather. The first hole in each pair should be made at the point where the taper ends and the widest part of the leather begins. The holes in each pair should be ½ inch (1.3 cm) apart for small and medium-sized birds, 1 inch (2.5 cm) apart for large birds, and 1½ inch (3.8 cm) apart for extra large birds.

3. Make a third pair of holes (C) at a distance from pair A that will allow a comfortable fit around your bird's leg. Typical distances are ½ inch (1.3 cm) for small birds, 1 inch (2.5 cm) for medium-sized birds, 1½ inch (3.8 cm) for large birds, and 2¼ inch (5.7 cm) for extra large birds. (This distance should be based on the diameter of your bird's leg, which might vary from these recommendations, depending on species and sex. Once you have a jess that works well, keep a pattern of it for future use.) The distance between the holes in pair C should be the same as in step 2.

4. Using a blade, make a slit between each pair of holes. To avoid cutting the leather past the holes and weakening it, start from each hole and cut towards the center.

Fig. 5.18
Modified traditional jess design by John Karger,
Last Chance Forever/The Bird of Prey Conservancy.

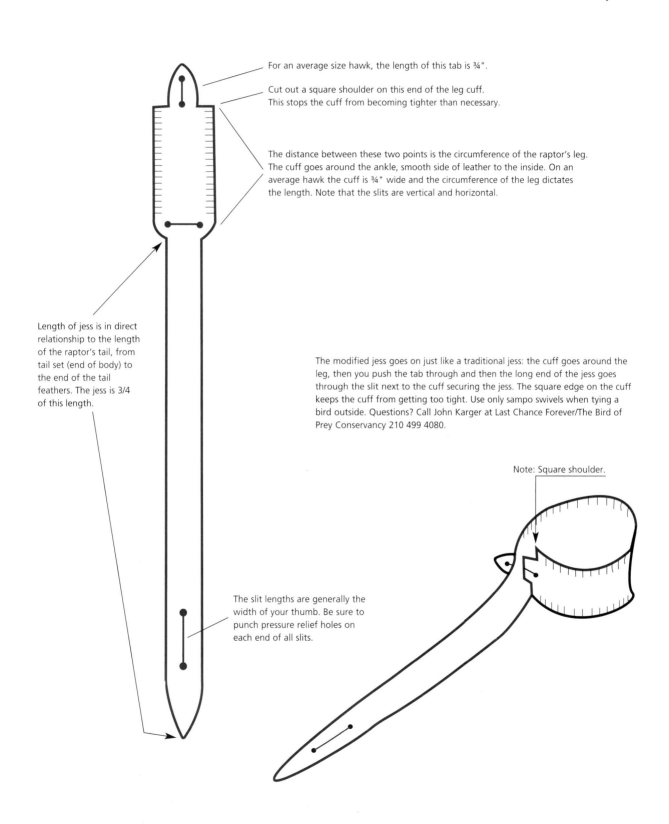

For an average size hawk, the length of this tab is ¾".

Cut out a square shoulder on this end of the leg cuff.
This stops the cuff from becoming tighter than necessary.

The distance between these two points is the circumference of the raptor's leg.
The cuff goes around the ankle, smooth side of leather to the inside. On an
average hawk the cuff is ¾" wide and the circumference of the leg dictates
the length. Note that the slits are vertical and horizontal.

Length of jess is in direct
relationship to the length
of the raptor's tail, from
tail set (end of body) to
the end of the tail
feathers. The jess is 3/4
of this length.

The modified jess goes on just like a traditional jess: the cuff goes around the
leg, then you push the tab through and then the long end of the jess goes
through the slit next to the cuff securing the jess. The square edge on the cuff
keeps the cuff from getting too tight. Use only sampo swivels when tying a
bird outside. Questions? Call John Karger at Last Chance Forever/The Bird of
Prey Conservancy 210 499 4080.

Note: Square shoulder.

The slit lengths are generally the
width of your thumb. Be sure to
punch pressure relief holes on
each end of all slits.

5. Using a blade, make small slits on the edges of the leather between slits A and C. This will add a little flexibility to the leather as it wraps around your bird's legs.

6. Soften the leather by massaging it with a leather conditioner.

Apply conventional traditional jesses: (figure 5.17b,c)

1. Wrap the end of the jess with the two slits around your bird's leg.

2. Insert the shorter end through the slit closest to your bird's leg, exposing the slit (figure 5.17b).

3. Pull end 2 through the slit exposed (figure 5.17c).

Instructions for making and applying the modified traditional design are provided in figure 5.18.

Jess extender

A jess extender is a 6–8 inch long (15.3-20.3 cm) piece of leather or nylon cord with a slit or loop at each end (figure 5.19). If your bird gets twisted or hog-tied when tethered, a jess extender attached between the jesses and swivel can help. It will relieve some of the tension formed when your bird repeatedly turns around and will prevent your bird's swivel from flipping in between the jesses (or through long fragile tail feathers). These are the two most common causes of twisted jesses and leashes. For larger birds, two leather jess extenders used back to back are often necessary to prevent leather breakage. To attach an extender to the jesses, follow these steps (figure 5.20).

1. Place both jess ends through one slit or loop of the extender.

2. Pass the other end of the extender through the jess slits and pull it through.

Leash

A leash can be made out of several types of material, but most handlers use leather, climbing rope, or nylon cord. The material chosen must be strong enough so that it won't break as your bird bates, flexible enough to be manipulated, and weather resistant. Sometimes, people get creative and braid the leash material to improve its strength and longevity. You can find instructions for making these fancy leashes in the selected readings at the end of the chapter.

Fig. 5.19
Jess extender. (Gail Buhl)

(a)
A leather jess extender.

(b)
A jess extender attached between the swivel and leash of a hawk.

Do not use nylon dog leashes, chains, or metal cables as bird leashes. The nylon dog leashes fray easily, and the chains and cables are heavy, difficult to manage, do not have the needed flexibility, and can easily cause injury to your bird.

Make a button leash: A leather or nylon leash is a rather easy piece of equipment to make. All it really needs is a "button" at one end large enough so that it won't pass through the swivel. If you're using a leather leash, you can make a button the same way as shown for the leather aylmeri jesses (figure 5.14). You can create buttons for nylon cord leashes by making a double knot on one end and then melting the tip so it doesn't unravel (make sure to melt and burn the leash ends in a well ventilated area). For both types of leashes, you can slip a washer, a piece of shrink tubing, or a circular piece of leather larger than the knot between the knot and the swivel as an extra safety measure, to make sure the knot won't slip through the swivel. If you decide to use a leather leash, it must be conditioned regularly with a softener and monitored carefully for wear.

Make a slit leash: Some people prefer rope or nylon leashes without a button. These can be made in the same fashion as the nylon jesses (figure 5.15). To attach a nylon leash, make a slipknot by feeding the looped end through the swivel and then the free end through the loop. These leashes are more secure than button leashes and as a result are more difficult to get off.

The length of your bird's leash will depend on its use. If you need a leash just for holding the bird on your glove, 3 feet (0.9 m) should suffice for most birds. If, on the other hand, you use your leash to tether your bird, the length will depend on both the size of the mew and the distance from the top of the perch to the ground (chapter 4, Housing). For a tethered bird, longer does not mean better. Allowing a tethered bird a large space to move around increases the probability of injury.

5.3 SENSITIVE SPECIES

There are a few species that present challenges when equipment is applied and they require additional maintenance and monitoring. They include vultures, owls, and hawks with feathered legs.

5.3a Vultures

Vultures have a variety of unique behaviors that help them survive in warm climates and defend themselves against threats. One such

Fig. 5.20
Applying a jess extender. (Jen Veith)

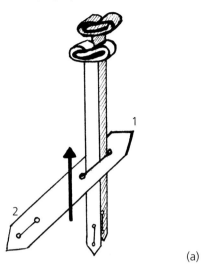

(a)
Place both jess ends through slit at end 1.

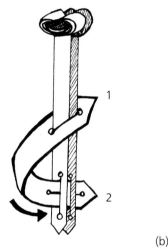

(b)
Slip end 2 through the slit in each jess and pull snugly.

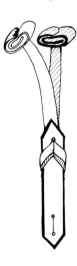

behavior is that they mute (urinate and defecate) on their legs to keep cool in warm climates. This can pose a real problem with anklets, jesses, and leashes. These pieces of equipment get soiled with urates and then malfunction, and cause skin irritations. There are several management approaches you can take to avoid or minimize this problem.

The safest situation is to free-loft a vulture without equipment. If your vulture is destined to be a program bird, this will only work well once the bird is trained to accept your approach and allows you to put equipment on without restraint. The modified aylmeri jess system works well for this because the anklets can be applied, removed, and cleaned easily. Traditional jesses can also be used in this fashion but it is not recommended to apply these if a bird is going to be tethered.

A second approach is to free-loft a vulture with anklets but no jesses. The anklets would need to be monitored daily and changed frequently to avoid mute build-up.

If your vulture cannot be free-lofted due to behaviors and/or training status, daily examination of the equipment is highly recommended. Providing ample bathing opportunities is essential to keeping a vulture's legs and equipment clean.

5.3b Birds with Feathered Legs

Many owl species and a few hawk/eagle species (rough-legged hawk, ferruginous hawk, and golden eagle) have feathers that extend down their legs to their feet. The presence of jesses may cause feather loss under the anklet portion and, if not monitored, skin abrasions. To avoid these problems, the anklets should be kept soft and must be applied loose enough to allow sufficient rotation of the leather around the leg. If you have problems with feather loss and/or leg abrasions, remove the jesses and anklets and free-loft your bird until feathers grow back. When you reapply equipment, make the anklet portion a little looser (but don't allow it to slip over the foot or above the hock joint).

5.4 EQUIPMENT FOR BIRDS TRAINED FOR PROGRAM USE

Although this book is not going to cover managing birds for free-flight programs, TRC though it prudent to mention that there are a few additional pieces of equipment you would need for this undertaking. These include a pair of flight jesses, a lure, and a transmitter/receiver system.

5.4a Flight Jesses

If you're going to free-fly your bird, you'll need a pair of "flight jesses" that you put on before each performance. A free-flying bird must wear aylmeri jesses that do not have slits at the ends (the slits that a swivel would pass through). If a bird wearing jesses with slits takes a detour and leaves for a while, branches and other obstacles could get caught in the slits, trapping it.

People who free-fly a bird often punch a tiny hole at the bottom of each jess through which they pass a clip swivel when the bird's flight is finished and the bird is held on the glove. A clip swivel has a clip on one end and a rotating ring on the other (figure 5.3). You can tether your bird with this swivel and avoid making an extra pair of jesses; however, clips can be faulty, so if you use a clip swivel, don't leave your bird tethered outside of an enclosure or inside an enclosure that has multiple birds tethered.

5.4b Lure

If you are free-flying your raptor, it might become "stubborn" or distracted and not want to fly down to your gloved hand, no matter how hard you plead. One tool many falconers use for recovering a raptor is a lure. A lure is basically any small object a bird has been trained to associate with food. Most often, it is designed to resemble a natural food item, for example, leather cut, shaped, and stuffed into a "bird" with extended wings (figure 5.21). A line is attached (leather or nylon) and food is clipped to the lure. The lure is then presented to a bird, either by swinging it to catch the bird's attention and then placing it on the ground, or by placing it on the ground and dragging it a bit.

A lure is a pretty simple item to make. If you don't want to make your own, however, you can purchase one from sources listed in appendix D.

Fig. 5.21
A typical lure used to train a flighted raptor so it can be recovered.
(Gail Buhl)

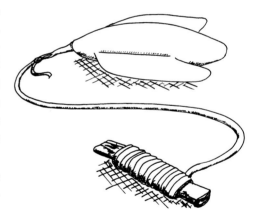

5.4c Transmitter/Receiver

Even the most dependable and best-trained bird may, on occasion, decide to take a detour during a free-flight session. Most raptors won't fly very far, usually remaining in view. Falcons, however, have a tendency to stretch their wings and cover quite an expanse of territory.

Most falconers and bird handlers apply a transmitter to their bird so they can locate it if it flies out of view. Transmitters can be mounted on a bird's leg, tail, back, or around its neck. These electronic devices deliver a specific frequency signal that can be picked up with a receiver and antenna. The entire electronic package can be quite expensive; however, you can use a receiver and antenna with as many transmitters as you want, as long as they all use frequencies within the receiver's detectable range. This is especially handy if you fly multiple birds throughout a program and don't have time to take a transmitter off one bird and place it on another. A list of transmitter suppliers is provided in appendix D.

5.5 SUMMARY

Having and maintaining the proper equipment is essential for keeping your bird as comfortable as possible, avoiding injury to your bird (and in some cases, to you), and preventing the loss of your bird. TRC strongly recommends investing in good, proven equipment and supplies; don't think you can get by with cheap, weak leather or makeshift leashes. If you use inadequate equipment, one day you won't "get by"; instead you'll say goodbye to your bird as it flies off or disappears, modeling its faulty equipment. If you aren't sure whether the items you've made or purchased are adequate, seek the advice of falconers or other raptor handlers.

5.6 SUGGESTED READINGS

Arizona Falconer's Association. 1990. *A Manual for the Apprentice Falconer.*

Beebe, F.L., and H. M. Webster. 2000. *North American Falconry and Hunting Hawks.* Denver, CO: North American Falconry and Hunting Hawks.

California Hawking Club. 2000. *Apprenticeship Manual.* California Hawking Club, Inc.

California Hawking Club. 1995. *Apprentice Study Guide.* California Hawking Club, Inc.

Kenward, R. 1987. *Wildlife Radio Tagging.* Academic Press.

Kimsey, B. and J. Hodge. 1992. *Falconry Equipment.* Houston: Kimsey/Hodge Publications.

Morgan, D. 2002. *Braiding Fine Leather.* Centreville, MD: Cornell Maritime Press, Inc.

Naisbitt, R. and P. Holz. 2004. *Captive Raptor Management and Rehabilitation.* Surrey, Canada: Hancock House.

Upton, R. 1991. *Falconry: Principles and Practice.* London: A & C Black.

Table 5.1 Recommended average sizes of equipment

Bird Size	Anklet Width	Jesses		Raw Aylmeri Jess Length	Grommet Size
		Strap Length	Width		
<200g (small)	¼" (0.6cm)	4" (10.2cm)	3/8" (0.9cm)	6" (15.2cm)	"00" – 3/16" (0.5 cm) diameter hole, or eyelet
200-900g (medium)	½" (1.3cm)	7" (17.8cm)	½" (1.3cm)	8-11" (20.3-27.9cm)	"0" - 1/4" (0.6cm) diameter hole
900-2000g (large)	¾" (1.9cm)	9" (22.9cm)	5/8" (1.6cm)	11-12" (27.9-30.5cm)	"1" – "2" 5/16-3/8" (0.8-0.9cm) diameter hole)
>2000g (extra large)	1-1 ¼" (2.5-3.2 cm)	10.5" (26.7cm)	¾-7/8" (1.9-2.2 cm)	15" (38.1cm)	"3" - 7/16" (1.1 cm) diameter hole

Chapter 6: MAINTENANCE CARE

Keeping a captive raptor looking good and feeling healthy takes work. Just as people undergo a variety of daily hygiene routines (combing hair, brushing teeth, trimming fingernails) captive raptors also need maintenance care. Your bird's overall physical appearance will reflect the quality of care and management you provide and can sometimes indicate medical problems as well. Weight loss, broken or unpreened feathers, overgrown beak and talons, or foot sores indicate medical, nutritional, and/or management problems that must be addressed immediately. As is highlighted in many chapters of this book, it is critical to maintain constant vigilance over the overall health and appearance of your bird, and to keep up-to-date records on all aspects of its care and behavior.

6.1 RECORD KEEPING

Paperwork, paperwork, paperwork. It seems as if many things in life inevitably lead to more paperwork. Well, possessing a raptor also creates quite a paper trail. It's extremely important, however, to keep well-organized and updated records on any bird in your possession.

As recommended in chapter 1, Permits, start a file on your bird. In that file include a maintenance-care schedule for checking the bird's weight, feathers, feet, beak and talons, equipment, and behavior. The Raptor Center uses the forms shown in figure 6.1. Sometimes health and maintenance factors and behavior patterns can be anticipated or explained from knowledge of this history.

For birds that are on display only, TRC recommends conducting general maintenance checks once a month, as long as the birds are observed daily for any signs of illness or injury, and good food consumption records are kept. Birds that are manned should be briefly checked every time they are on the glove, the most important considerations being the bird's weight, foot condition, and equipment condition. Then, a more in-depth assessment can be made once a month.

ED BIRD INSPECTION SHEET

Name _____ Handler _____ Location _____

Species _____ Target Weight Summer _____ Winter _____ Highest _____ Lowest _____

Write additional comments on back of sheet

NW = Item Needs Work   OK = Item Fine

JAN Date/Wt.	Beak	Feet	Fthrs.	Gear

FEB Date/Wt.	Beak	Feet	Fthrs.	Gear

MAR Date/Wt.	Beak	Feet	Fthrs.	Gear

S

APR Date/Wt.	Beak	Feet	Fthrs.	Gear

MAY Date/Wt.	Beak	Feet	Fthrs.	Gear

JUN Date/Wt.	Beak	Feet	Fthrs.	Gear

JUL Date/Wt.	Beak	Feet	Fthrs.	Gear

AUG Date/Wt.	Beak	Feet	Fthrs.	Gear

SEP Date/Wt.	Beak	Feet	Fthrs.	Gear

OCT Date/Wt.	Beak	Feet	Fthrs.	Gear

W

NOV Date/Wt.	Beak	Feet	Fthrs.	Gear

DEC Date/Wt.	Beak	Feet	Fthrs.	Gear

Fig. 6.1
Forms used by TRC to keep track of health checks and maintenance care of education birds.

(a)
Daily observations.

(b)
Monthly summary.

Morning Courtyard Rounds
Week Of: _____

	Monday	Tuesday	Wednesday	Thursday	Friday	Saturday	Sunday
Mew 1 BAEA "Leuc"							
Food left (type/amount)							
√ Tether (length, twisted?)							
Comments							
Mew 2 PEFA "Artemis"							
Food left (type/amount)							
√Tether (length, twisted?)							
Comments							
Mew 3 MERL "Taiga"							
Food left (type/amount)							
Heating element on?							
Comments							
Mew 4 BNOW "Whisper"							
Food left (type/amount)							
Heating element on?							
Comments							

6.2 GENERAL COMMENTS

There are a few guidelines you can follow to make the handling and maintenance exam less stressful on you and your avian educator:

1. If you must cast (grab) a manned raptor (for veterinary care, beak maintenance, etc.), it's wise to have the bird cast by someone other than the primary handler. This prevents the bird from associating this less-than-positive experience with the "friendly" handler.

2. If your raptor is manned, TRC recommends grabbing it directly off the glove rather than out of its enclosure. This will not only help minimize the risk of injuring the bird (especially if you are a novice bird handler), but will also keep the enclosure a safe haven. Keep in mind that even the best-trained raptor might show aggressive or defensive tendencies when cast. Don't worry; most raptors are "forgiving" and will stand on your gloved hand again after the exam.

3. When your bird is restrained, cover its head with a towel or hood (chapter 5, Equipment). Also, try to keep the area as quiet as possible and request minimal activity near the head of the bird. These tips will help keep the bird calm, reduce struggling, and make the experience less stressful for both you and your bird.

4. Do not feed your bird on the day of an exam until after the exam has been completed. If you need to cast it for a beak coping, etc., you don't want it to have food in its stomach or crop — it may regurgitate and aspirate into its lungs.

5. When capturing a trained raptor, some handlers ask their bird to step onto a "special" glove that differs from the regular handling glove. They either wear this glove when they enter the mew or have an assistant wear the glove and transfer the bird to him/her before it is cast off the glove. This way, the bird will not associate the regular handling glove with the act of being grabbed; the handling glove will remain a safe place to perch. When entering the enclosure, this special glove should be hidden behind the handler's back so the bird does not instantly get suspicious and react.

6. An exam should be conducted in a room that is different from where your education programs are presented. Once

again, you don't want the bird to associate negative experiences (being grabbed) with the location where it is expected to be calm and feel secure.

7. Do not have the daily caretakers capture display birds. It is wise to have someone who has little to do with them be the "bad guy." This will help prevent the birds from associating negative experiences with their caretakers and keep them calmer during daily feedings, cleanings, etc.

8. TRC recommends conducting exams indoors; however, if the exam must be conducted outside, make sure it is done during the coolest part of the day to prevent your bird from overheating. Also make sure it is conducted in a secure location to prevent your bird from accidentally escaping during the procedure.

6.3 HANDLING EQUIPMENT

Regardless of what species of raptor you have, there are several things you'll need for handling it during exams and other procedures. Most important is a pair of gloves to protect your hands and arms from sharp wandering beaks and talons. For small birds, such as American kestrels, Eastern and Western screech owls, and Northern saw-whet owls, short leather or sheepskin gloves that extend just above the wrist are nice, and can be purchased at hardware stores, garden shops, building-supply stores, and most department stores. Welder's gloves ($5 to $10 per pair at most liquidation or home improvement stores) work well for medium-sized, large, and extra-large birds such as red-shouldered hawks, great horned owls, vultures, and eagles.

If you prefer fitted gloves, a variety of sources make custom gloves or produce a few different sizes (appendix D). However, do not use a glove that comes with a clip attached. When holding a bird on the fist, the clip is handy for attaching to the swivel to prevent escape. When casting your bird, however, the clip can get in the way and potentially injure your bird.

In addition to welder's gloves, for eagles, ospreys, owls, and vultures (who have a nasty habit of biting and twisting skin), TRC strongly recommends a welder's jacket to protect your upper arms and chest. You can make such a jacket out of soft leather or suede, or buy it from a welding supply store.

If your bird is strictly a display animal and not trained to the glove, you'll probably have to grab it from a free-lofted position. If this is the case, we recommend wearing protective eyewear. Most birds, except for some imprints and some members of more aggressive species such as barred owls and Swainson's hawks, won't deliberately attack you. Accidents do happen, however, and it's better to be prepared by protecting your eyes.

The last pieces of equipment you'll need are a padded table, a quiet space in which to work, a spray bottle to mist your bird if it gets overheated, and a few towels or a hood (appendix D). These things will help keep your bird calm and reduce struggling during the procedure.

6.4 HANDLING A BIRD FOR MAINTENANCE OR PHYSCAL EXAMS

Whenever you perform a maintenance exam or treatment on your bird, you'll need to use some form of restraint to ensure that the procedure can be done quickly and efficiently, and that the bird won't injure itself, the handler, or the examiner. If your bird is manned, you can easily do most maintenance checks with it on your gloved hand, and no further instructions are necessary (see chapter 8, Training). If your bird is used strictly for display, however, and not trained to the glove, you'll need to use a different method of restraint. The goal is to place your raptor on its back or vertically against the handler's chest and restrain it firmly and safely for the duration of the procedure. Sometimes you'll need to cast your manned raptor too. The methods described below can be used for both manned and unmanned birds.

First, just a few clarifications of terms used in the following paragraphs. Handling your raptor for an exam involves four major steps: capture, control, restraint, and recovery.

The capture involves encouraging a bird into a position in which it can be grabbed (or "cast"), such as a corner of a display enclosure or for a manned bird, a gloved hand, and then physically removing it from the perched or standing position. Once the capture is completed, the raptor will probably struggle and flap with vigor. The handler will then need to gain control of the bird. The restraint is next, requiring secure positioning of the bird on a table or in the restrainer's arms so the procedure can be performed. The last phase is the recovery, or returning the bird to a perched position on a gloved hand or an upright position in its enclosure.

6.4a Capture

Trained raptors

If your raptor is well trained, the capture will involve stepping the bird onto your glove in the usual manner, transporting it from its housing area to the exam station, and then gently but firmly casting it so it can be restrained. Do not cast the bird in its enclosure. Remember that your bird's housing environment is a "safe" zone and nothing negative should happen there.

TRC recommends working with a partner who will be responsible for grabbing the bird and restraining it while the exam or procedure is completed. Small raptors, such as American kestrels, Eastern and Western screech owls, and Northern saw-whet owls, should only be grabbed using a body-grab technique.

With the bird perched on your glove, have your partner approach it from behind (so if the bird bates it will bate away from you) and quickly and smoothly encircle both hands around the wings, lifting the bird off your glove at the same time. A towel can also be placed over the bird before it is grabbed (figure 6.2).

TRC does not recommend grabbing small birds using the leg grab technique defined below. Grabbing a small bird by the legs can cause injury, especially if only one leg is secured and/or the bird is grabbed by its relatively thin lower legs.

Medium and large raptors (short-eared owls, broad-winged hawks, great horned owls, red-tailed hawks, etc.) can also be grabbed off the fist using a body-grab technique, although in the case of large raptors, it is more difficult with a bird weighing over 1,000 g (35.3 oz). The leg-grab method is more often used for medium, large, and extra-large raptors.

With the bird perched on your gloved hand, have your partner approach it (from either the front or the rear), making sure not to show the bird the gloves until the last possible moment. Then, in one swift and decisive movement, your partner should grab the bird's legs up close to the bird's body (the legs are quite muscular here) one leg in each hand (figure 6.3a).

Turning the lights off in the room may make this task easier, as the bird will be calmer during the grabber's approach. However, only experienced handlers should grab a bird in the dark.

Many people prefer to have a secondary handler not only take the bird out of the mew, but also cast it off their fist themselves. To be safe doing this, a glove should be placed on the handler's free hand. However, this glove should be worn only after the bird has been recov-

Fig. 6.2
Grabbing a small manned raptor using a body-grab-with-a-towel technique.
(Gail Buhl)

ered from the mew. If the handler enters the mew with both hands gloved, the bird will learn what will follow and will be more difficult to capture, especially if it is free-lofted.

Once the second glove is applied, the handler should reach up from in front of the bird and grasp both legs such that his/her pointer finger is between the legs and his/her thumb and third finger are on the outside of the legs. Then, the handler should lift the bird off the fist (if it has not already jumped), turn the back of the bird to the ground and gain control as described below for a leg grab (figure 6.4).

Some people choose to grab both legs from behind the bird, but TRC finds this a little more awkward and it forces the handler to lean into the bird, potentially putting him/her in a more vulnerable position for getting bit or footed. It also risks damaging the tail feathers. Do not cast an extra-large raptor (such as an eagle) off your fist by yourself. Two hands are needed to cast an extra-large raptor safely.

Extra-large raptors should be grabbed with a leg-grab technique. Keep in mind, however, that these birds are very strong. Bald eagles not only flap vigorously, but also have the habit of biting in defense. Also, remember that extra-large birds have large wingspans. To prevent injury to your bird, make sure that after the grab, the bird's wings don't hit the floor or any objects. That requires a large, obstacle-free space. Grabbing extra-large birds off the fist in the dark is not recommended — the grabber needs to see where the feet and beak are at all times to prevent getting seriously injured.

Untrained raptors

If your raptor is a display-only bird and not glove trained, the capture will be a little more difficult, especially if the bird is flighted. One method is to encourage the free-lofted bird to move into a corner where it can be grabbed. If the enclosure is large and the bird flighted, you might need two people to corner the bird. Usually, when you enter the enclosure a flighted bird will fly back and forth. If you hold your arms as high as possible and wave them, the bird will eventually fly to the ground.

Once the bird is in a reachable position, small unmanned birds should be grabbed with the body-grab technique mentioned above; medium and large raptors can be grabbed by either a leg grab or a body grab using a towel (figure 6.5a). If you want to use a body grab, make sure the towel completely covers the bird's head. From behind, use your arms to pin the bird's wings to its body, position your hands such that the bird's thighs are between the 2nd and 3rd fingers of each hand, lift the bird, and hold it into your chest (figure

Fig. 6.3
Grabbing and controlling a trained raptor from a fisted position.
(Ron Winch)

(a)
Grasp one leg (above the hock joint) in each hand.
(sequence continued on page 151)

Figure 6.4
A single-handed, single person approach to grabbing a bird off the glove.
(Ron Winch)

Fig. 6.5
Grabbing and controlling an untrained raptor using a body-grab technique. (Gail Buhl)

(a)
From behind, cover the entire bird with a towel or blanket.

(b)
Grasp one thigh in each hand, pinning the wings to the body with your arm and lift the bird to your chest.

(c)
Rearrange your hands to control both feet with one hand and hold in a vertical or cradle position.

6.5b). Quickly rearrange your hands such that your strongest hand is controlling the legs as described below and watch out for the bird's beak. You are extremely vulnerable holding a bird in this position, especially if the towel comes off the bird's head. Once you have control of the legs, gently but firmly take a front hold of the bird's neck to control its head and/or place a thick towel or glove behind its head so it won't bite your chest.

One note of caution if you want to use a towel: If the bird grabs and holds on to the towel, do not rip its talons free. This could cause a talon sheath to come off, which is a very painful and bloody injury (chapter 7, Medical Care).

Another method of capture sometimes used for small and medium-sized raptors involves using a fishing net with small holes. This technique is a little risky as the bird can be hurt if it is accidentally hit by the rim of the net during pursuit, and feather damage can occur if the bird gets tangled and is difficult to remove. Do not attempt to use a net for large or extra-large raptors — it is dangerous for both the bird and the handler.

When your educator is being captured, it might exhibit some species-specific defensive behaviors, so be prepared. Red-tailed hawks, for example, might spread their wings and sit on their tail, ready to strike with their feet; great horned owls often elevate their body feathers and hold their wings in a convex position, stomping back and forth to be intimidating; and vultures vomit their putrid stomach contents.

6.4b Gaining Control

You can be certain that your raptor (trained or not) will attempt to free itself of your grip following a grab. Your mission is to quickly gain control of the bird so it doesn't hurt itself (or you) and is held in a comfortable position. Keep in mind that different species respond differently following a grab.

Small raptors are quite squirmy and often attempt to bite as you gain control. Larger raptors are more powerful. Golden eagles, for example, tend to bite less than bald eagles, but their feet are much stronger and should always be grasped firmly. Falcons bite hard and turkey vultures not only bite and twist skin, but also have the tendency to regurgitate their stomach contents, practically knocking out the grabber with the foul smell — a pretty neat defensive strategy. This just proves that you should be prepared for almost anything.

Leg-grab technique
Once the bird has been grabbed off a glove or lifted off the ground

using the leg-grab technique, it will flap vigorously. The bird should be rotated downward so that its back faces the ground, but the handler should not bend over and let the bird get closer to the ground (figure 6.3b). This action will usually cause the bird to reduce its flapping. Be careful not to let the bird's wings hit any objects, including the ground. Also, the grabber should make sure to keep the bird's beak away from his/her arms and legs.

Then, the grabber should rearrange his/her hands so that one hand restrains both legs as follows:

- With palm facing down, the index finger should be placed between the bird's legs, and the thumb and third finger around the outside of the legs to secure a grip (figure 6.3c).

- The legs should be held close to the feet to minimize the bird's ability to strike out and foot someone.

- Then, the grabber should take his/her free hand and reach under the bird's back to fold in its wings, being careful not to grab the flight feathers in the process.

- Medium, large, and extra-large raptors can then be held in the cradle of the grabber's arm (figure 6.3d) or vertically against his/her chest. However, if holding a bird against a chest, it is recommended to either place a thick towel or glove behind its head or instruct the grabber to grasp the bird's neck from the front so it can't turn around and bite.

Body grab technique

Gaining control of a medium-sized bird that has been body-grabbed is similar. With the bird pressed against the handler's chest (its head must be covered), his/her hands should be rearranged as described above so that both legs are held with one hand. Then, either the bird should be rotated so it rests in the cradle of the grabber's arm (figure 6.5c) or maintained in the vertical position against the grabber's chest.

6.4c Restraint

Once the bird is under control and in a secure grasp, it should be laid on its back on an exam table or held up against the handler's body. For safety and ease of handling, TRC most often prefers to place a bird on an exam table.

When the bird is on the table, the grabber should remove his/her hand from around the bird's body and slowly place this hand (or index finger, if the bird is small) across the bird's chest and under its

Figure 6.3 *continued*

(b)
Turn the bird so that its back faces the ground.

(c)
Reposition your hands so that one hand (palm down) holds both legs near the feet.

(d)
Reach under the bird's back to collect its wings.

Fig. 6.6
Techniques for restraining a raptor during exams.

(a)
Restraining a raptor on its back on a table. (Ron Winch)

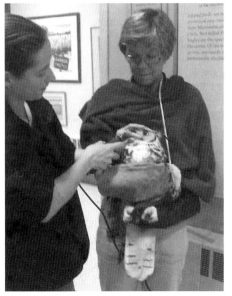

(b)
Restraining a raptor in a vertical position.

lower beak so it can't bite the person restraining it or the examiner (figure 6.6a). At the same time, the hand securing the beak should be able to hold in the wing farthest from the restrainer; the restrainer's thumb, arm, or body should be used to hold in the wing closest to him/her. The restrainer should make sure not to press hard on the bird's chest or its ability to breathe will be compromised. A gentle but firm hold will suffice to prevent the bird from flapping or biting the examiner.

One last tip: in order for the restrainer to stay on friendly terms with his/her coworkers, it is important for that person to let the examiner know if his/her grasp is loosened or lost. If you and your partner don't communicate in this fashion, the bird may injure one of you with its feet or beak.

To restrain a bird against your chest, gently rotate the bird from the cradle position to a vertical position against your chest and move the arm not holding the bird's legs so it reaches over the bird's chest and restrains both wings (figure 6.6b) At this point it is strongly advised to have your coworker place a thick towel or glove behind the bird's head, so it can't turn around and bite your chest. It is also a good idea to slide a large or extra-large bird down a little so its beak is farther away from your face. This may be easier to do if the handler sits down.

6.4d Recovery

Once the procedure is completed, you'll want to recover the bird by either repositioning it on the primary handler's gloved hand or, if it isn't trained, returning it directly to its enclosure. Do not let a trained bird jump up from the table to the gloved hand. Instead, to remove the bird from the table, the restrainer should slide the bird towards the edge of the table, and then back into the cradle of his/her arm, hand (for a small raptor). Then, to reposition a trained raptor, have the primary handler take the jesses in one hand (see chapter 8, Training) and the restrainer lift the bird forward onto the glove. Sometimes a raptor will immediately bate, but a well-trained bird should come back onto the handler's glove.

If your bird is hot and excessively stressed, make sure it is calm and cooled down (it might pant or gular flutter for a few minutes) before it is returned to its housing. Some species, such as peregrine falcons, gyrfalcons, and golden eagles, tend to overheat rapidly after being restrained. Have a spray bottle ready to lightly mist such a bird to facilitate its recovery.

To recover an untrained raptor, the bird should be carried back to its enclosure and, with all doors closed, placed gently on the floor on its back. As the handler removes his/her grasp, the bird should be gently rolled away from the handler. In this fashion, the bird will continue to right itself and gain its composure before flying to a perch or walking up a ramp.

Oftentimes, if a raptor is placed directly on its feet on the floor, it immediately flaps to get away and hits its wrists on the ground, causing injury. TRC doesn't recommend setting an untrained raptor on a perch, for it will inevitably try to fly away from the handler and might land on the enclosure walls or ceiling, again increasing the risk of injury. Keep an eye on the bird to make sure it recovers (stops panting or gular fluttering). This is especially important on hot days, when the stress of handling can lead to excessive heat production.

6.5 THE MAINTENANCE EXAM

As mentioned earlier, you should regularly check your bird's weight, feathers, beak, talons, feet, and equipment to make sure the bird looks and feels healthy. TRC recommends doing complete maintenance checks once a month, but for birds that are out regularly for programs, a quick check every time the bird is on the glove will help prevent problems from developing. You can't check your bird too often.

6.5a Weight

You can learn a great deal about the mental and physical health of a captive raptor by its weight. Often, when a bird gets sick, it either loses its appetite or loses weight despite eating. A responsible manager, therefore, should not only weigh each bird regularly but also keep detailed records of the type and amount of food offered and consumed daily.

Seasonal changes can trigger changes in a bird's weight. In the northern latitudes, the fall migratory urge can signal a raptor to deposit fat to either fuel its anticipated journey or store it for the cold days ahead. The bird's appetite increases, and so should its daily food ration (see chapter 3, Diet). During the winter, a large raptor (red-tailed hawk, great horned owl) might gain up to 200 grams (7.1oz). In the spring, when temperatures and day length increase, a raptor may put itself on a diet, often losing all the additional winter weight and leaving some food uneaten. Be aware of these natural weight fluctuations and keep records of them for future reference. However, if you are ever in doubt about what con-

Fig. 6.7
Weighing an unmanned raptor on
a scale. (Ron Winch)

(a)

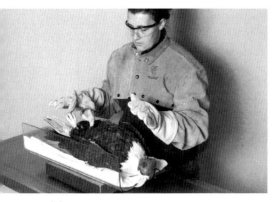

(b)

stitutes a healthy weight for your raptor, don't procrastinate. Call your veterinarian or an experienced caretaker immediately.

How to weigh a bird

There are several different techniques you can use to weigh a raptor, depending on whether the bird is trained or untrained. As explained in chapter 8, Training, a well-trained raptor will step onto a scale that has been modified to include a perch and allow its weight to be recorded easily (figure 8.10). Unfortunately, untrained raptors are not quite as cooperative. Display birds can be captured and placed on an appropriate scale to get a weight, trained to freely enter a crate that can then be weighed, or trained to step up onto a perch fastened to a scale.

If a display bird needs to be captured to record its weight, follow the techniques recommended in the beginning of this chapter to capture, grab, and gain control of it. Then, it can be placed on a scale as follows:

Small raptors: Wrap the bird in a small towel and place it keel down on the scale. The towel will prevent the bird from wiggling out and generally, small birds are less active if placed on their keels instead of on their backs when wrapped. You will need to let go of the bird completely to record an accurate weight. Make sure to either zero the scale with the towel on it before wrapping the bird or weigh the towel afterwards to get the correct bird weight.

Medium-sized, large, and extra-large raptors: Set medium, large, and extra-large birds on their backs on a scale without a perch. This works best if a large towel is present on the scale and opened such that once the bird is on its back the towel can be folded over to cover the bird's head.

Place the bird on its back; very slowly remove your hand from around the bird's body (but keep your leg hold), and fold the towel over the bird's head. Take this hand and either hold it at your side or place it horizontally above the bird's head. Then — again, slowly — remove your hold on the bird's legs until your hand is completely free (figure 6.7). Read the weight and regain your grasp on the bird's legs.

Even though you may be anxious about regaining a grip on your bird, don't move quickly when grasping the legs and body of your raptor. Fast, jerky movements are unsettling to a bird and will cause it to try to foot you or to jump off the scale. Always move slowly. Regain your grasp of the legs first, remove the towel from the bird's head, and slide the bird toward you and into the cradle of your arm.

That might sound easy, but you can be assured that not all

untrained raptors will lie on their backs without a fight. Here are some general tips on weighing and handling that can be used even with a "problem child":

- Always weigh any bird, either trained or untrained, indoors. This will prevent the potential nightmare of watching your bird leap off the scale and zip into the nearest tree, perhaps with equipment on.

- Keep human activity around the weighing area to a minimum during the procedure. Lots of activity will make an untrained raptor edgy and less likely to stay on the scale long enough to record its weight.

- Make sure to have enough free space around the scale that the bird won't injure itself if it does bate or jump off.

- To calm a particularly jumpy bird, darken the room, apply a hood, or wrap the untrained raptor in a towel covering its eyes before weighing (just make sure you deduct the weight of the towel).

- Pull down the window shades or find some other way to cover any windows. If your bird gets loose, it might see the window as an escape route and crash into it, injuring itself or even fatally breaking its neck.

A display bird can also be weighed by a variety of training techniques. One method is to train it to enter a pre-weighed crate. Food is often a good motivator for training raptors (chapter 8, Training) and a display bird can be trained to enter a crate for a food "treat." The crate and bird can then be weighed without the experience being negative. A second method is to train a bird to step onto a scale in its display. This works especially well with vultures and ground-dwelling species.

For consistency, always weigh your bird on the same scale and at the same time of day, so it is at the same point in its feeding/digestion cycle. Digital scales are especially nice and can be purchased from a variety of sources (appendix D). Scales with one- or two-gram (0.04–0.07 oz) increments are nice to use for small raptors. For additional information on scales, see chapter 5, Equipment.

How often to weigh your bird
Every time a manned raptor is on the glove, it should be weighed.

Not only does this provide valuable information about the health of your bird, it also helps to keep the bird trained so it easily steps on and off of the scale. TRC recommends weighing program birds a minimum of once per week. A display bird can be weighed once per month once it is acclimated to a display and is known to be eating regularly. However, daily food consumption records must be kept and if there is any question about a bird's nutritional intake or health status, the bird should be weighed.

6.5b Feathers

Birds are unique in many ways, but their most obvious defining feature is feathers. These beautiful outgrowths of skin enable a bird to fly, serve as insulators to protect the bird from heat and cold, protect the tender skin from injuries, and function in behavioral displays and camouflage. You might be thinking that this is fascinating information, but why is it included in a book about captive raptor management? Well, feathers also function as indicators of a raptor's health and the adequacy of its management.

Poor diet, improper housing, poor handling skills, and weight control during the molting season can all lead to weak and damaged feathers. Even if you don't fly the bird for audiences, any raptor you handle or display should be kept in good feather condition; it reflects the bird's overall health and the care you provide.

Before potential feather problems are described, some general information about raptor feathers will be presented. Understanding the basics about feathers (anatomy, types, number, tracts, molting seasons, and patterns) will help you understand your bird's behavior and determine whether any problems exist.

Feather anatomy
A feather consists of many different parts, as illustrated in figure 6.8.

- Inferior umbiculus: the opening through which the feather receives blood and nourishment when growing

- Calamus (quill): the round, semitransparent portion between the inferior umbilicus and the beginning of the vane

- Rachis: the main feather shaft

- Vane (web): made up of closely spaced barbs that have barbules. These barbules have rolled edges and tiny hooklets that interlock the barbs together. When your bird preens, it is hooking loose barbs to form a smooth vane.

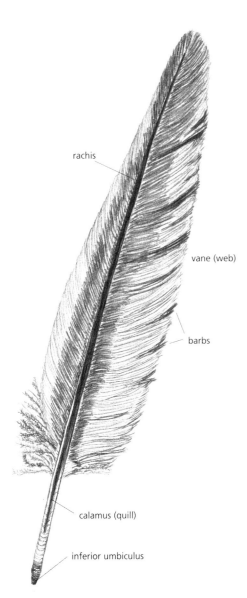

rachis

vane (web)

barbs

calamus (quill)

inferior umbiculus

Fig. 6.8
Structure of a contour feather.
(Gail Buhl)

Types of feathers

There are several different types of feathers covering your bird's body (figure 6.9). Each type varies in its location and function:

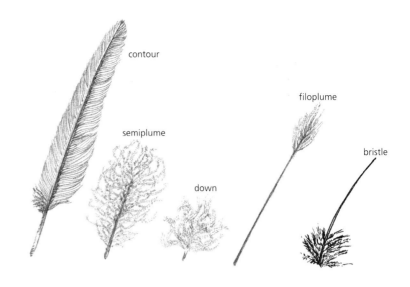

Fig. 6.9
Different kinds of feathers: contour, semiplume, down, filoplume, bristle.
(Gail Buhl)

Contour feathers: Contour feathers have vanes on both sides of the shaft and are moved by a series of muscles attached to the walls of the follicle (a depression in the skin from which a feather grows). Flight and tail feathers are contour feathers.

Semiplume feathers: Semiplume feathers lack barbules and hooklets on their barbs, so the vanes are fluffy and loose. They are commonly found under contour feathers, providing insulation and flexibility for the movement of the contours.

Down feathers: Down feathers lack a main shaft, barbules, and hooklets on their barbs. They are located next to the skin and serve mainly to insulate.

Bristles: Bristles have a stiff rachis but few barbs. They are commonly located around the eyes, nostrils, and mouth. Great horned owls have bristles on both sides of their beak and common barn owls have bristles on their toes. These feathers serve a tactile function, allowing a bird to feel things close to its mouth or feet.

Filoplume feathers: Filoplume feathers lack barbs on a majority of the shaft, except at the tip, and have sensitive nerve endings in the follicles. They surround contour feathers, especially at the nape and back, and may play a role in feather movement in these areas.

Number of feathers: Although the total number of feathers varies with each species, the numbers of major flight and tail feathers are similar for most birds of prey. The numbers we provide are "normals"; every once in a while you will run across an individual or species with an extra flight feather here or there.

Raptors generally have ten primary flight feathers and twelve to sixteen secondary flight feathers (figure 6.10). The flight feathers provide forward momentum, while the secondary feathers provide most of the lift. Most raptors have twelve tail feathers, which are essential for steering, providing lift, and maintaining equilibrium in flight. Species that make sharp, quick turns (accipiters) have rela-

Fig. 6.10
Raptor flight and tail feathers
identified using a common
numbering system. (Gail Buhl)

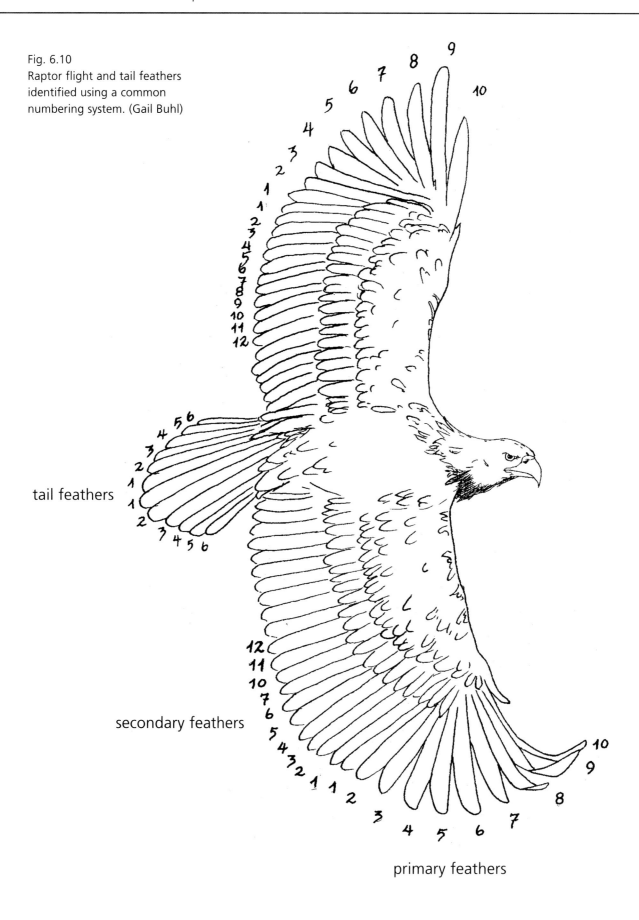

tail feathers

secondary feathers

primary feathers

tively long tails. Tail feathers are very delicate and break easily with inappropriate management and handling.

Feather tracts: Contrary to what many people think, raptors don't have feathers growing out of their entire bodies. In fact, feathers are located in specific tracts and overlap to give that fully feathered look. Therefore, when you part the feathers, you'll see areas that appear bare.

Molting seasons and patterns: In the wild, raptors follow a pretty regular molting season and pattern. Molting, which is the process of feather replacement, usually occurs between April and September in the Midwestern states. Many species begin their molt in April, while others, such as the great gray owl, begin in middle to late summer. Adult raptors generally molt once per year and don't necessarily molt every feather. It takes about four to eight weeks for one feather to grow in completely, so the entire process is gradual. Raptors lose only a few feathers at a time, so their activity isn't impaired in the wild.

Feather replacement usually occurs in a pattern, with certain feathers replaced before others. Many raptors replace the innermost primaries (those at the wrist) first and then work out to the wing tip (#1–10). Falcons begin with primary #4 and then molt in both directions. Secondary feathers are typically molted beginning at both ends (the wrist and elbow) and working toward each other (figure 6.10). The tail is replaced from the center out. Molting also tends to be bilaterally symmetrical; the same feather on each wing is replaced at the same time. This helps maintain symmetry in flight.

Captive raptors may undergo a modified molt, influenced by factors such as daylight, temperature, weight control, and other stresses. These factors can all delay the beginning of a molt, slow down its progress, stop it altogether, cause entire tails to molt at once, or cause new feathers to develop weak spots (called stress marks). Typically, a bird housed outdoors goes through a standard molt unless it is sick, stressed, its weight is dropped too quickly for flying purposes, or it is exposed to a new living area. A raptor housed indoors, on the other hand, is more likely to have an abnormal molt.

To ensure a normal healthy molt, make sure your bird is exposed to a day length and temperature comparable to those outside, and that it isn't excessively stressed during the molting season. It is also critical to have your bird on a healthy diet that includes a variety of prey items. Many caretakers increase the amount of food offered during the molt to meet the increased ener-

Fig. 6.11
Feather damage.

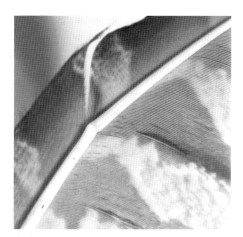

(a1)
Bent feather. (Ron Winch)

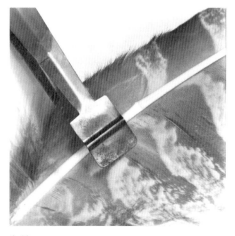

(a2)
Bent feather being repaired with a feather straightener. (Ron Winch)

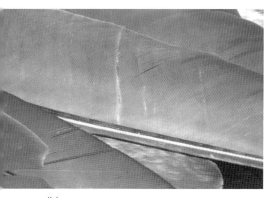

(b)
Stress marks

gy demand of the molting process. However, if you do this, monitor your bird's weight to make sure it does not get too fat. Obesity can lead to serious health problems (chapter 7, Medical Care).

Feather problems: If a raptor is managed properly, its feathers should remain healthy and molt normally. If your management plan is inadequate, feather damage can occur. Broken, bent, weak, parasite-infested, damaged, pinched, unpreened, or plucked feathers all indicate problems (figure 6.11).

Broken feathers: Broken feathers most often result from improper housing, poor handling, or inadequate means of transport. Cage material, perch placement, and means of housing (free-lofted or tethered) will all influence the condition of your bird's feathers (see chapter 4, Housing).

Your bird should be housed so that it can fully open its wings without touching any object. If the bird is antsy and has a tendency to grab on to the side of its caging, it will undoubtedly break its tail feathers. A bird exhibiting this type of behavior may need to be tethered in its mew either permanently, or during the part of the year when it is most active (breeding season, for example). Perches should be placed so that your bird's tail and wing tips don't touch the ground or the sides of the mew.

Sometimes a raptor will break feathers if it isn't handled properly. This occurs most frequently with birds that are on display and captured periodically for routine physicals. A manned bird, however, can also break feathers when handled, especially if its wings hit an object when it bates, an inexperienced handler tries to remove it from its mew, or it bates in such a fashion that its jesses and leash pass through its central tail feathers. Whenever you handle any bird, be conscious of its feathers. Chapter 8, Training, offers tips for proper handling of both manned and display birds.

Another situation that can lead to broken feathers is transport. Unless a bird rides comfortably in its carrier, it can easily break feathers. It is important to use a transport carrier appropriate for the species and to make adjustments as your bird dictates. Chapter 9, Transporting Raptors, offers guidelines for transporting your bird and providing the most appropriate carrier.

Bent feathers: Bent feathers can result from improper housing or handling. If a feather shaft bends but does not break, it can be straightened (figure 6.11a). A feather straightener is a tool available from Northwoods Limited (appendix D). It is a specially designed metal "iron" that, when heated and applied, softens and straightens a feather shaft. Other options are to take a wet rag, heat it in the microwave until hot, and then press and slowly slide it over the bent shaft area, or dip the feather in a cup of warm water.

Weak feathers: Weak feathers sometimes occur during a molt when a bird's feathers develop stress marks, hunger traces, or fault bars. All three terms describe the same condition: the vane of the feather has damaged barbs and thus a weakened area (figure 6.11b). Feathers grow relatively rapidly and any interruption to the blood supply of a growing feather (even if the interruption is short lived — a day or less) can cause a stress mark.

Stressors such as sickness or injury, inappropriate diet, movement into new housing, or training can cause growth interruption. Weak feathers can also develop if a bird's weight is decreased for free-flight demonstrations too rapidly during a molt. Keep these contributing factors in mind as you manage your bird, to prevent poor feather development. Feathers with stress marks break much more easily than healthy feathers.

Parasites: Parasites can also cause feather problems. Chapter 7, Medical Care, describes different kinds of parasites found on raptors. Certain types of parasites, primarily lice and mites, feed on feathers, newly developing and mature, causing damage to the feather vanes (figure 6.11c). Lice are large enough to be seen crawling in and on feathers; mites are small and are usually found after damaged feathers are noticed. These parasites chew the feather vanes, especially around the head and neck. If you see lice crawling on your bird or notice feathers that look chewed, dust or spray your bird with an appropriate compound (appendix D) or contact your veterinarian.

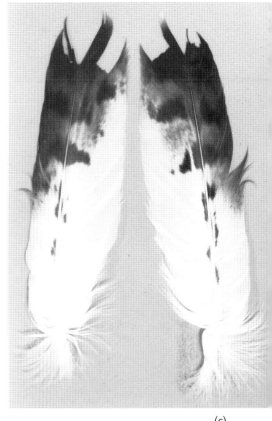

(c)
Chewed feathers. (Ron Winch)

Pinched feathers: Pinched feathers are growing feathers that are closed off at the base. This stops the growth and the feathers usually come out. Sometimes, a raptor pinches off what appears to be a normally developing feather (figure 6.11d). Causes for this can include a parasitic irritation of the feather follicle (mites present inside the developing shaft), excessive stress (nutritional or other), or a viral infection (see chapter 7, Medical Care). Watch your bird carefully. If your bird appears "off" and/or continues to do this, contact your veterinarian for assistance.

(d)
Pinched feather. (Ron Winch)

Damaged blood feathers: Damaged blood feathers occur during a feather's growth period. During this time, blood is present in the shaft, providing the feather with nourishment. Each newly develop-

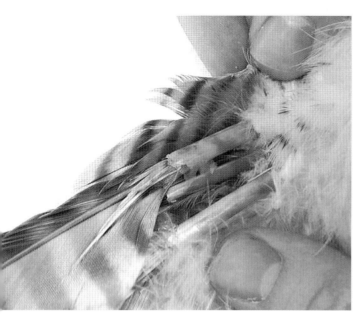

Fig. 6.12
Healthy blood feathers.

ing feather, called a blood feather, is very delicate, and the follicle from which it grows is very sensitive (figure 6.12). Damage to a blood feather can cause excessive bleeding and appears to be painful. Raptors with some permanent wing injuries are prone to damaging blood feathers because of their lack of full wing use.

If your bird damages or breaks a blood feather, you can stop the bleeding by properly restraining the bird, as described earlier in this chapter, and applying pressure to the damaged area with gauze or a clean towel. The bleeding should stop as the blood clots, but this may take a little while. If the bleeding doesn't stop, or damage to the feather is severe and painful, contact your veterinarian for treatment suggestions. Such a feather can be pulled out, but it's a painful process that is best done under anesthesia, and must be done carefully to avoid permanently damaging the follicle. If the follicle is damaged, the feather might not grow back or might grow crooked.

Unpreened feathers: Unpreened feathers are another sign of problems. A raptor that is healthy and comfortable in its surroundings will preen its feathers to keep them well groomed and organized. Sometimes, however, when a bird stops preening, its feathers become ruffled and dirty, and the outer sheath from developing feathers isn't removed. Often, if a bird is sick, has permanent nerve damage, has damage to its preen gland (uropygial gland), is under constant stress, or doesn't have access to a bath pan, it won't preen normally.

Lack of preening means that you need to examine your bird to make sure it isn't sick, as well as provide it with what it needs to be comfortable. Sometimes misting a bird's feathers or perching it outside with a pan of fresh water and moderate sun exposure will stimulate it to preen.

Feather plucking: Feather plucking is a condition often encountered by psitticine (parrot) owners, so it is not really thought about with raptors. Sometimes birds pluck their feathers due to a medical condition such as a parasitic or viral infection. Other times, however, the cause is behavioral. In parrots, boredom, lack of attention, poor environmental conditions, a new owner, excessive dampness, or a new stimulus in the environment (such as another bird or pet, a new child, rearrangement of furniture, new surroundings, etc.) can all cause feather plucking behavior. The birds engage in excessive preening to the point where they pull out their body feathers, leaving themselves bald in severe cases.

This condition also sometimes occurs with human-imprinted raptors (see chapter 2, Selecting a Bird for Education). These birds are socialized to people and without suitable interaction, routines, and a constant environment they can turn to feather plucking activity for stimulation. Stopping this behavior requires a medical assessment to rule out medical causes and then an evaluation of the overall management of the bird. Some change needs to be made; either a slower transition to a new handler or environment, increased enrichment to keep the bird stimulated (chapter 8, Training), or maybe even a transfer to a different facility.

Imping damaged feathers

Feathers that are plucked, pinched, or damaged while in blood cannot be manually replaced. You will have to wait for your bird to molt new feathers. However, feathers that are broken, severely bent, chewed on by parasites, or weakened severely by stress bars can be manually replaced by a process called imping. Imping is an age-old technique first developed by falconers and perfected over time to the efficient practice used today. Basically, a new feather is fitted into the bird's original feather shaft and secured with glue. This feather will then be naturally replaced when a bird goes through its normal molt.

There are two methods often used for imping feathers, the shaft method and the needle method. The former uses a piece of shaft from another feather to connect the new feather to the old shaft still on the bird. The needle method involves inserting a lightweight rod (bamboo is the material of choice) into the shaft of the new feather and the shaft remaining on the bird to connect them. TRC prefers the needle method since you don't need to come up with additional feather shafts, glue holds strongly to the bamboo, and it is easier to fit bamboo into different sized shafts by whittling. Thus, this is the method described below.

Supplies: TRC recommends saving molted feathers from your bird (or, if you are at a rehabilitation facility, feathers from raptors that have died) so that if you ever need to replace any, you'll have a supply. To protect the feathers, store them in an envelope in an area safe from moths (a freezer works well). Naphthalene (moth balls) has the potential to be toxic to birds so it is not recommended to store feathers in this fashion. Label the feathers with the species; age of bird; side of body (left wing, right wing, or tail) and feather number, if known (refer to figure 6.10 for the conventional numbering system).

In addition to feathers, you'll need several pieces of equipment for the imping process (figure 6.13a):

- A sharp blade

- Bamboo or chopsticks for medium, large, and extra-large sized birds; or guitar string for smaller birds.

- Nail trimmers

- Five-minute epoxy (a 50/50 mix of epoxy resin and hardener is best; you can also use superglue, but it doesn't give you as much time to align the feathers after you glue them)

- Small strips of paper

- Paper clips

Procedure: Once you've gathered this equipment, you're ready to begin. The first step is to prepare the new feather(s) before you have your bird restrained:

1. Make a list of the feathers that need to be replaced (figure 6.13b).

Fig. 6.13
(all photos by Ron Winch)

(a)
Feather imping equipment.

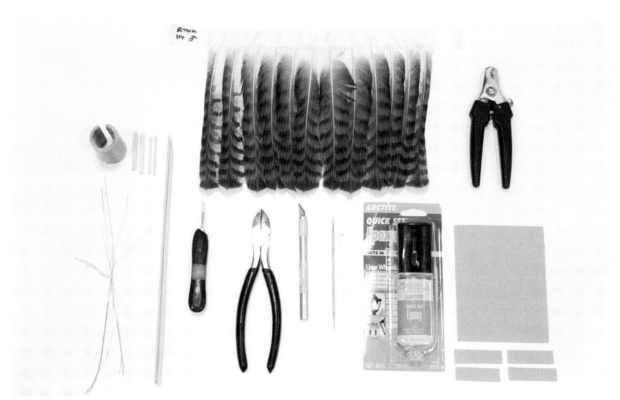

Top, from left to right: Replacement feathers, nail trimmer.
Bottom, left to right: Material for imping needles (bamboo, guitar string), whittling tool, cutting tool, tools for cleaning inside of feather shafts (dental instrument, paper clips), glue, paper.

2. Try to select feathers identical to the ones being replaced. Feathers should be from the same species, and if possible, the same age and sex as the bird being imped. Within a species, the size of feathers differs between the sexes and ages and you want to make sure that the overall size of the feathers chosen are similar to the size of other feathers on your bird. In addition, the individual feathers chosen must be the same feathers that need to be replaced; a number 9 primary should be replaced with a number 9 primary.

3. Clean out the feather shafts with a paper clip or other blunt object.

4. Prepare imping needles. Bamboo makes good needles because it is sturdy and lightweight. For each needle you will need about two inches (5.1 cm) of bamboo. To start, whittle about three-quarters to one inch (1.9-2.5 cm) of bamboo so it fits snugly into the new feather shaft (use the shorter length for smaller birds). Leave the other inch or so at a larger diameter. A little additional whittling might be necessary when the feather is fitted (figure 6.13c). If using guitar string, make sure the diameter string you use fits snugly into the shaft and roughen it up a bit with a fingernail file so the glue will adhere.

5. Set everything up before you begin working on your bird. Lay out new feathers with imping needles inserted. Have epoxy, trimmer, paper, etc., handy.

6. If you need to replace more than a few feathers, TRC recommends that the bird be anesthetized to minimize stress. Please consult your veterinarian for assistance.

7. Now you're ready to begin the actual imping process. With your bird either anesthetized or properly restrained on a table, follow the next steps:

8. Fit new feathers into the bird. If more than one feather needs to be replaced, start with the innermost one. This will allow you to rest each new feather against either an original feather or a newly fitted one to get the orientation correct. Then, using the nail trimmer cut off the broken feather so that at least one inch (2.5cm) of shaft remains (do not use a scissor to cut the feather as this will cause the shaft to split). This is the area where you will make the implant (figure 6.13d).

Fig.6.13 (b)
Broken tail feathers that need to be replaced on a young red-tailed hawk.

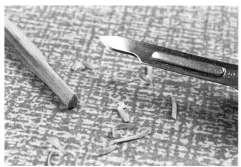

(c1)

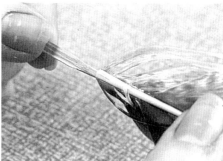

(c2)

Imping preparation. Half of an imping needle is whittled to fit snuggly in the shaft of a replacement feather.

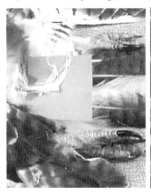

(d)
Implant site.

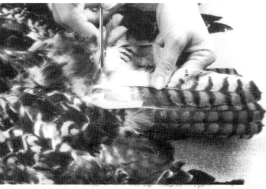

(e)
With the bird restrained, a new feather is trimmed to the appropriate length.

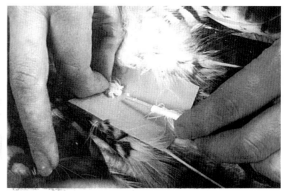

(f)
The second half of the imping needle is whittled so it fits snuggly in the shaft remaining on the bird.

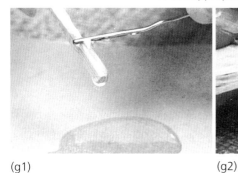

(g1)

(g2)

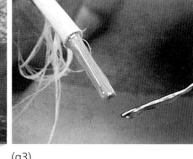

(g3)

Five-minute epoxy is applied first to the end of the needle fitting into the replacement feather, and then to the end sliding into the shaft remaining on the bird.

(g4)

(h) Completion of the imping process.

(You will notice that a feather shaft gets larger as it approaches the skin so imping at this site will be easier and the connection point will be more protected by the body. If you imp near a feather tip or at mid-length, the new feather will be more likely to break.). Then, using the other wing or half of tail as a reference, adjust the new feather to the appropriate length by trimming the shaft at the bottom. (figure 6.13e). Make sure that the shaft of this new feather smoothly meets the shaft remaining on your bird. If you trim the new feather and are using bamboo needles, you might need to whittle the bamboo a little more so that it still fits. Then, whittle the bamboo sticking out of the new feather so it fits snugly in the shaft remaining on your bird (figure 6.13f). Repeat this process with all the feathers to be replaced, before gluing.

9. Place strips of paper under each cut shaft on your bird so that when you glue in the new feathers, glue won't touch the surrounding feathers.

10. Glue the new feathers in. First, apply glue to the part of the bamboo or guitar string that fits into the new feather and insert it. Then apply glue to the rest of the bamboo or string and slide it into the shaft on your bird, making sure not to get glue on any of the feathers (figure 6.13g 1-4). Before the glue dries, make sure to orient the feathers so they sit properly against the others and remove the paper strips.

11. Let your bird recover completely from the stress or the anesthesia before returning it to its enclosure.

6.5c Feet

One of the most sensitive parts of your captive raptor is its feet. In the wild, a raptor maintains healthy skin on the bottom of its feet by perching on a variety of surfaces, keeping its feet moist by bathing and exposing them to rain, and maintaining a proper varied diet full of vitamins and minerals. Once the raptor is under your care, you control all of these factors.

TRC cannot stress enough the importance of providing your bird with varied perches of proper sizes and surfaces, appropriate housing with weathering areas, and adequate nutrition. Without these, your bird's feet will ultimately develop sores that could become infected and potentially life-threatening. Keep a hawk's eye on the condition of your bird's feet. What may have worked well for years may change, based on a bird's age, activity level, weight, and climatic conditions.

As part of your maintenance check — or anytime you have reason to handle your bird on the glove or for a physical exam — develop the habit of checking its feet. For a trained program bird this can be done by slowly lifting up the middle and innermost toe to reveal the foot pad underneath (figure 6.14a). This can either be done by the handler or by an assistant. Many times a manned bird is more cooperative with the handler than with another person in front of it. If your bird is a biter, you can use a prop, such as a narrow dowel or rod, to lift up the toes.

For a display bird that is not trained, checking the feet will be a little more complicated. The bird will have to be cast and restrained while an assistant examines the feet. To open the toes on a restrained bird, the handler should stretch out the legs to relax the extensor tendon in the feet. Then, it will be easier for the examiner to separate the toes by pulling back on the third and rear toes (figure 6.14b). If detected and treated early, any soreness or wounds on the bottom of the feet (bumblefoot) can be treated effectively, preventing more serious foot problems from developing. For more information on detecting and treating bumblefoot, see chapter 7, Medical Care.

6.5d Beaks and Talons

If you've ever dreamed of becoming a beautician (and even if you haven't) you'll be interested to know that the raptors in your care need a little "cosmetic" assistance with their beaks and talons every now and then. In the wild, raptors "freak" (or "feak") their beaks on rough surfaces, wearing down the sides to maintain an appropriate length and sharpness. They keep their talons manicured by exposing them to various textured surfaces.

In captivity, this self-maintenance often breaks down, leading to overgrown or cracked beaks (figure 6.15) and talons. Long beaks can make it difficult for a bird to tear apart its food and can even lead to beak breakage. Overgrown talons can lead to bumblefoot (see chapter 7, Medical Care). It's your responsibility to make sure that the beak and talons of a bird in your care are well manicured.

Coping is a falconry term used to describe the process of reshaping beaks and trimming talons. To cope your bird's beak, you'll need to know what the natural shape and length of the beak are; these vary with species. Refer to other birds of the same species (if you have access to them) or to photographs of wild birds as models (figure 6.16).

Fig. 6.14
Techniques for examining a raptor's feet.

(a)
Checking feet with bird on the glove.

(b)
Checking feet with bird restrained on exam table.

Coping beaks

This procedure can be done efficiently and with minimal discomfort to your bird if you use the proper equipment and technique.

Equipment: The best equipment for beak coping is a Dremel® tool with an appropriately sized bit. TRC recommends using a variable speed tool. Which speed to use depends on skill and personal preference. TRC has found that on a small bird, it's very easy to take off too much length or narrow the beak too much if you use the higher speeds. Slower speeds are also nicer for coping falcon beaks, with their delicate notches.

Rotary bits are made of several different types of grit and come in several different sizes. The bits TRC uses are made of silicon carbide. For smaller birds, a 1/8-inch (0.32 cm) diameter cone-shaped bit is used, for larger birds, a 1/4-inch (0.64 cm) bit. To cope falcon notches, a 1/8-inch (0.32 cm) cylindrical bit works well (figure 6.17). You can buy Dremel tools and bits in building supply stores, hardware stores, or the tool section of some department stores.

It is also a good idea to have some Kwik Stop® handy in case bleeding occurs. This product is available at pet stores, drug stores, or through your veterinarian.

Procedure: As a general guideline, your bird's beak might need a little shortening and reshaping every two to three months. If you're not familiar with the process of beak coping, seek the assistance of a local falconer or experienced rehabilitator or veterinarian. When you're ready to perform the procedure, make sure you have all the needed supplies handy.

First, restrain the bird on its back or in an upright position as described earlier in this chapter. Gently and quickly control its head. There are two main methods used by handlers to do this. One involves grabbing the head from behind. To do this, slide your hand, with the palm up, around the back of your bird's head so that its head fits between your index and third fingers. We can almost guarantee that when you do this, the bird will try to bite. Distracting the bird with your other hand as you grab its head might make the task easier. Wearing a small glove can also prevent you from getting bit but it often makes the bird more nervous and it is more difficult to gain a secure grip of the head and mouth with a gloved hand. Prop the bird's mouth open by placing your thumb in the soft corner of the mouth (figure 6.18a).

The other method for restraining the bird's head is as follows: with the bird facing you, slowly move your hand to the side or top of the bird's head and grasp its head with your thumb under the lower beak and your other fingers on top of the bird's head. Watch

Fig. 6.15
Overgrown beaks

Fig. 6.16
Normal beak shapes

Fig. 6.15 a
Red-tailed hawk

Fig. 6.16 a
Red-tailed Hawk

Fig. 6.15 b
Great horned owl

Fig. 6.16 b
Great horned owl

6.15 c
Peregrine falcon

Fig. 6.16 c
Peregrine falcon

Fig. 6.15 d
Bald eagle

Fig. 6.16 d
Bald eagle

Fig. 6.17
Recommended rotary bits for coping raptor beaks. Left to right, bit for small beaks, medium to extra-large beaks, and falcon notches. (Ron Winch)

Fig. 6.18
Head hold for coping beaks.
(Gail Buhl)

(a1)
Method one: With your palm facing up, gently grasp the bird's head from behind.

(a2)
Prop the bird's mouth open by placing your thumb in the soft corner of the bird's mouth.

(b)
Method two: With the bird facing you, slowly move your hand to the bird's side and gently grasp the head using your thumb to secure the lower beak and your other fingers to secure the top of the head. Prop the bird's mouth open by placing your index finger in the soft corner of its mouth. (Jen Veith)

the bird closely and be careful not to get bit. Then, slide your pointer finger into the soft corner of the bird's mouth (figure 6.18b).

Next, once the bird's head is firmly under control, start coping with the Dremel tool on low speed, at least until you feel comfortable with the procedure. Beaks grow in both length and thickness, so you should run the tool over the sides of the beak to thin the sides and edges and remove layering (figure 6.19). If the beak is really long, you can shorten it with a Dremel or by taking a nail trimmer and clipping off the tip to the desired length. However, take off small bits at a time. As beaks grow, so does their blood supply and if you are not careful, you can hit the quick and cause bleeding. If you nick the end of the quick, minimal bleeding will occur; if you cut into the quick, more extensive bleeding will occur.

Once you have reached the desired length, take the Dremel and shape the beak tip so it comes to a sharp point. Smooth over any cracks. Shorten the lower beak if necessary so that the upper and lower beaks close comfortably (figure 6.20a). Smooth over the edges of the lower beak so that they don't grow and curve inward (figure 6.20b) and smooth the sides to eliminate layering (figure 6.20c). Be careful not to injure the bird's tongue. Hold it down with the finger you are already using to keep the bird's mouth open.

One of the trickiest aspects of beak coping is positioning the notches in falcons. The upper beak has one notch on each side and the bottom of each notch should rest at the middle of the lower beak when the beak is closed (figure 6.16c). Beaks grow from a growth plate near the cere (fleshy area above the beak) and as the beak grows in length, the notches move farther and farther down (figure 6.15c). They need to be pushed back up to the proper position (fig-

Fig. 6.19
Coping the upper beak of a bald eagle.
(Ron Winch)

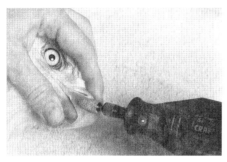

(a) Thinning the sides.

(b) Reshaping the edges.

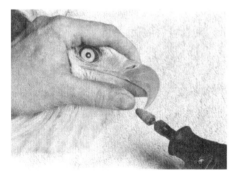

(c) Shortening the tip.

(d) Sharpening the tip.

ure 6.21). By doing so, you will increase the beak tip length which will then need to be shortened.

For falcons, it is recommended to move the notches into their correct position before trimming the beak length. The lower beak has ridges that the notches rest in and if the notches are not kept in the proper position, the ridges can grow improperly too. If you are coping a falcon's beak and notice large cracks extending vertically from the notches, remove the cracks entirely, even if that requires removing the notches. If left alone, even small cracks can turn into larger cracks, and major beak damage can occur. Don't fret; your falcon won't look like a red-tail for long; the notches do grow back.

If the beak starts to bleed during coping, stop immediately. You have reached the beak's blood supply and shouldn't cope any further. Apply pressure to the damaged area for a few minutes, using a piece of gauze or a tissue (don't use cotton; it tends to stick to the beak and often starts the bleeding again when you try to remove it). If the bleeding is stubborn, either apply some Kwik Stop with a cotton-tipped applicator, or put silver nitrate (from a silver nitrate stick) on it. If the bleeding persists, call your veterinarian.

There are a few cautionary notes to make for new "rapticians":

First, make sure to protect both your eyes and that of your avian partner. Beak dust is light and tends to "fly" into the air, potentially getting into your eyes or the eyes of your bird. Not only is it irritating, but think about what a raptor touches with its beak – beak dust is full of bacterial organisms that could cause an eye infection. Therefore, it is a good idea for the person coping to wear protective eyewear and when coping is completed, to flush your bird's eyes with a few drops of saline (warm water will do if you don't have access to saline).

Second, if you have long hair, pull it back to prevent it from getting caught in the Dremel tool. It is amazing how quickly hair gets wrapped around a rotary bit. In the same vein, be careful not to catch any of your bird's head or neck feathers in the rotating part of the tool. This will create a large tear in the bird's skin that will require veterinary attention (usually stitches). This has most frequently occurred with owl species that have dense feathers around their face. It is a good idea to have an assistant press down feathers around the beak while you are coping.

Tips for maintaining beaks

There are several things you can do to help your raptor maintain a healthy, well-formed beak:

• Watch to see where your bird usually eats and cleans its beak. Often, birds have a "favorite" spot, on a certain perch, rock, etc. Place a piece of moderately coarse sandpaper on or around a small section of the spot. Once your bird finds it, it may use it regularly. Just replace the sandpaper when it gets worn.

- Some falcon managers provide their free-lofted bird with a cement feeding block. The bird eats on the block and then feaks its beak on the cement to clean and maintain it. However, if a bird is tethered, we do not recommend leaving a cement block in reach, as consistent perching on this surface may cause bumblefoot.

- Provide your raptor with a whole-prey diet, including the heads of large prey animals (chapter 3, Diet). As it picks meat off the bones or tears off pieces, the bird will rub its beak against the bones, helping to file it. However, do not offer a head that is so small that your bird could swallow it whole and potentially get it lodged in its esophagus or crop.

- Cope your bird's beak regularly; don't let it get long and thick. Coping it into a natural shape is difficult when a beak is terribly overgrown.

Coping talons

Like coping beaks, coping talons can be done efficiently and with little discomfort if you use the right equipment and technique. Talons also grow in length and thickness so you must pay attention to both (figure 6.22).

Equipment: Coping talons requires only a few materials. For small and medium-sized raptor species, a cat nail-trimmer works well. For large and extra-large raptors, a dog nail trimmer or Dremel tool can be used. For nail trimmers, TRC prefers rounded end trimmers that open completely to surround a talon, not guillotine-style trimmers. You'll also need some Kwik Stop or silver nitrate sticks in case a talon bleeds.

Procedure: Keep talons a little dull and at an appropriate length by trimming the tips regularly. Most captive raptors require a trim about once a month. If using a nail trimmer, only trim off about one-sixteenth to one-eighth inch (0.16-0.32 cm) at a time. If you think the talons are still too long, continue to trim them, taking off very small amounts so you don't hit the blood supply. If using a Dremel, only take off a small amount; it is easy to get carried away and take off too much, leaving your bird with abnormally short and blunt talons. A Dremel tends to cauterize the vein if it is hit, so

Fig. 6.20
Coping the lower beak of a bald eagle.
(Ron Winch)

(a) Shortening the lower beak.

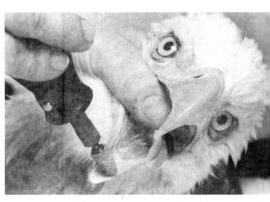

(b) Lowering the side edges.

(c) Removing side layering.

Fig. 6.21
Moving a falcon's notches into proper position. (Ron Winch)

(a) Shortening the sides if necessary.

(b) Moving the notches up.

Fig. 6.22
Normal length/shape of talons.

(a) Peregrine falcon.

bleeding will be less. Note that the talon on the middle toe (third digit) has a sharp ridge on one side. This aids in preening and should not be removed.

If your bird develops excessively thick talons, the extra layers on the side can be removed by a variety of techniques. One method of removal requires filing with a Dremel tool. If using a Dremel, you must be careful to smooth the talons evenly, create the appropriate shape, and avoid the blood supply. Some birds will not let you easily Dremel their talons without anesthesia. If this is the case, consult your veterinarian for assistance.

Another method for thinning talons is to apply wet ball bandages to the feet for twenty-four hours. The moisture softens the outer layers of keratin on the talons and they flake off, revealing a nice thin, sharp talon underneath. Properly applying these bandages requires training and if you are considering this technique, you should consult your veterinarian or an experienced raptor rehabilitator for assistance.

Lost talon sheath

One other concern regarding raptor talons is loss of a talon sheath. Once in a while, a raptor might get a talon stuck on a rope perch, some other part of its enclosure, or a towel/blanket when being grabbed. As it tries to free itself, the outer talon sheath may loosen and come off. This often causes substantial bleeding, and you can be certain that it hurts (just like losing a fingernail). It is critical to stop the bleeding, treat the talon bone, and protect it to prevent infection (chapter 7, Medical Care).

6.5e Equipment

If your raptor is manned, it's very important to check the condition of its equipment on a regular basis (see chapter 5, Equipment). Even the best leather will wear in time and need to be replaced. It's much better to notice this during a regular check than during a time of panic, when your bird is sitting in a tree with broken jesses.

It's a good idea to record when a piece of equipment is applied, and how it looks each time you weigh your bird. Usually, jesses break at the slit through which the swivel passes, and anklets twist and wear right next to the grommet. Also, if you tether your bird, examine its leash to make sure it isn't frayed, worn, or otherwise weakening. One last piece of equipment to check is the swivel. As swivels age, they sometimes stick and the rings don't rotate anymore. If your bird is tethered, this could be a serious problem, causing your bird to get tangled and "hog-tied."

If you ever doubt the security of the equipment your bird is wearing, replace it. The value of a pair of jesses is nothing compared to the value of your avian educator. You won't be sorry.

6.5g Behavior Patterns

Like people, birds can be very temperamental. They react differently to different people and can be very moody. Time of year, weather patterns, handling consistency, physical health, and imprinting on humans are a few factors that can alter a bird's behavior patterns. It's very important to keep detailed records of any changes in your bird's behavior and the circumstances surrounding them. This will help you to know when a behavior is normal for the bird, and when there truly is a problem.

6.6 SUMMARY

Conducting and recording regular maintenance exams on your raptor is extremely important. You control many aspects of your bird's existence, and you must make sure the bird looks and feels healthy. Maintenance exams should include checks on your bird's weight, feathers, feet, beak and talons, equipment, and behavior. Observing these aspects of your bird will help you catch potential health and management problems and correct them before they become serious. You will also be able to predict problems and behaviors based on past records. Preventing management problems is often much easier than solving them and will keep your bird looking and feeling healthy and well cared for.

6.7 SUGGESTED READINGS

Dunning, J.B. Body Weights of 686 Species of North American Birds, Monolith 1, *North American Bird Bander,* 1984. Phoenix, AZ.

Faaborg, John, and Susan B. Chaplin. 1988. *Ornithology, An Ecological Approach,* pp. 14-22. New Jersey: Prentice Hall.

Fox, N. 1995. *Understanding the Bird of Prey.* Surrey, Canada: Hancock House Publishers.

Naisbitt, R. and Peter Holz, DVM. 2004. *Captive Raptor Management and Rehabilitation.* Surrey, Canada: Hancock House Publishers.

Welty, J. C. 1982. *The Life of Birds,* 3rd ed. Philadelphia: Saunders College Publishing.

(b) Red-tailed hawk.

(c) Great horned owl.

(d) Bald eagle.

Chapter 7: MEDICAL CARE

Everyone agrees that preventing medical problems is much easier than curing them. Certain diseases might be unavoidable, but many illnesses and injuries can be prevented, or at least detected and treated early. Medical problems usually occur at extremely inconvenient times, cause major setbacks in schedules and lifestyles, and can be physically and emotionally painful. Your bird won't be able to solve its own medical problems; you agreed to be the problem-bearer and problem-solver in the team when you agreed to have the raptor under your care.

As we all know, there's no foolproof method for avoiding life's problems. Nevertheless, if you become knowledgeable about keeping a raptor in captivity, manage your bird appropriately, and prepare yourself for potential medical problems, the difficulties you and your bird experience will be fewer and less severe. Here are some helpful hints to get you started:

- Make sure that any bird you acquire is appropriate for you. Pick a species that is consistent with your experience level and can be housed in your climate. Learn about any species predispositions to diseases (for example, golden eagles are prone to aspergillosis, gyrfalcons to aspergillosis and malaria) and the medical history of the individual bird you have in mind (chapter 2, Selecting a Bird for Education).

- Before you get a bird, choose a veterinarian with whom you feel comfortable, and who can be the bird's primary medical caregiver. Make sure the veterinarian is willing and interested in caring for your animal. There are a few books available on medical care for raptors; you should make sure the veterinarian has access to them (see suggested readings at the end of this chapter).

- Research proper housing before constructing an enclosure for your bird. Proper size, materials, and shelter will avoid numerous injuries that can turn into medical nightmares (chapter 4, Housing).

- Find suppliers who can provide a variety of food items, and establish a good rapport with them. Make sure the food is free from chemicals, has been raised with proper nutrition, has been humanely and safely euthanized, and has been stored appropriately (chapter 3, Diet). Never feed questionable food. Game animals that have been shot should be avoided (they could contain lead), as should animals that have died from unknown causes.

- Gather the basic medical supplies listed in appendix E and store them where they will be readily accessible. When taking a bird off site, take an emergency kit with you and be trained on how to use it (appendix E).

- Once you acquire a bird, take the time to observe it for management problems and conduct a basic maintenance check regularly (chapter 6, Maintenance Care). This will help you detect early signs of some common medical problems and treat them before they become more serious.

Despite even the most diligent care, captive raptors sometimes experience illness and injury. If your bird is sick, it should be taken off display and not participate in programs until it is healthy. The most common medical problems fall into one of four categories (listed alphabetically, not in order of prevalence):

- Infectious diseases
- Management-related injuries and illnesses
- Nutritional disorders
- Parasitic infections.

Familiarize yourself with the medical conditions listed below so you can detect problems early and get them treated.

7.1 DISEASES

Tiny organisms such as bacteria, fungi, and viruses, can cause a variety of illnesses in your bird. If the cause of sickness is diagnosed quickly, many of the diseases can be treated and/or the bird medically supported while the infection runs its course. However, if the organisms go undetected and untreated, many of them can prove to be fatal. So, once again it is critical to keep a close eye on your bird and its health. As you will see, some of the raptor diseases listed are classified as zoonotic; this means there is the potential for transfer of the disease causing organism from a raptor to a person, although reported incidence of this is rare. Also keep in mind that veterinary medicine is dynamic and new treatments are being discovered all the time. Always consult with your veterinarian for the most up-to-date information.

7.1a Bacterial Infections

Avian tuberculosis (TB)

Avian TB is caused by the bacterium Mycobacterium avium. This is a different species of bacterium than what causes human TB (Mycobacterium tuberculosis). All birds are susceptible to avian TB and it seems to be more prevalent in captive birds (poultry, quail, pheasants, parrots, and some species of birds of prey) in northern states. Avian TB is a chronic disease, with few symptoms early on. As it progresses, however, it can cause respiratory problems, diarrhea, an increased appetite with accompanying weight loss, and lameness. It is difficult to diagnose, but liver biopsies are often performed to look for the offending organisms. In raptors, this disease is spread through either ingestion of infected prey or mutes, or inhalation of airborne bacteria. Therefore, it can spread easily between birds housed together or in close proximity.

Although it has rarely been reported, humans can contract avian TB. Respiratory infections and swelling of the lymph nodes below the jaws occur. Humans have a high resistance to the bacteria, and people who have contracted the disease were already ill with lung disorders and/or were immuno-suppressed. As with most diseases, the spread of avian TB can be prevented by thorough sanitation of your bird's living quarters and proper personal hygiene. In addition, if you feed poultry to your raptor, make sure that it comes from a reputable source and is clean (no mutes on its feathers) before presenting it to your bird.

Salmonellosis

Salmonellosis is a disease caused by bacteria of the genus Salmonella. These bacteria are often spread in situations where the environment has been contaminated (sewage waste water in urban areas, for example), or where large numbers of birds coexist in a small area (poultry farms, pigeon lofts, bird feeders, rehabilitation facilities, etc.). In raptors, the bacteria can be spread by the ingestion of either contaminated mutes or salmonella carriers (rats, mice, chickens).

Symptoms in captive raptors include sudden lethargy, inability or unwillingness to stand, decreased appetite, and diarrhea. If you diagnose the infection early, successful treatment involves supportive fluid therapy and antibiotics (metronidazole, 50mg/kg twice a day or 100mg/kg once per day). A raptor acutely infected (perhaps from eating contaminated food) but properly treated will usually recover and resume normal activity within a week. If you suspect your bird has a salmonella infection, contact your veterinarian immediately.

Salmonella bacteria can also be present in raptor pellets. If you or your organization supplies schools or other people with pellets, the pellets should first be sterilized. People should also be informed to wash their hands well after handling pellets to prevent a salmonella infection.

In people, salmonella infections are often referred to as "food poisoning" and if you have experienced this, you will know that the symptoms are unpleasant. Fever, vomiting, and severe diarrhea result and last one to three days.

To prevent this disease in both you and your raptor, make sure that your food source is "clean," keep the mew clean and disinfected regularly, and follow good personal hygiene practices after handling your bird, its pellets, cleaning its mew, etc. You could pick up this disease through contact with your bird's infected mutes, pellets, or contaminated food.

Tetanus

Tetanus is a disease caused by the bacterium Clostridium tetanii and is transmitted through breaks in the skin from penetrating wounds. Raptors can get the disease from being footed by another bird. They can also carry the clostridium on their feet and if a handler gets footed, there is the possibility of bacterial transfer into the wound. Therefore, people working with raptors MUST have an up-to-date tetanus immunization.

7.1b Fungal Infections

Aspergillosis

Aspergillosis is a disease that should get a red star. Whenever wild raptors are brought into captivity, get sick, or have their management plan changed, they get highly stressed and are at high risk for contracting this disease. Aspergillosis ("asper") is a potentially fatal fungal disease that most commonly infects the respiratory system in a wide variety of animals (figure 7.1). In fact, wild birds kept in captivity are more susceptible than birds in the wild for reasons discussed below.

Asper is caused by fungi of the genus Aspergillus. Aspergillus fumigatus is the most common species. The spores of this fungus are widespread in the environment and airborne spores usually enter the body through inhalation. In a healthy animal, the body's defense mechanisms are sufficient to prevent spore germination. Infective spores are more likely to germinate and cause disease when the body is unable to resist them, such as when the number of atmospheric spores becomes extremely high or the individual's immune system is suppressed due to constant exposure to stressors.

Captive raptors, which often experience at least a low level of stress, are highly susceptible to this disease. Food, temperature, crowding, and fear are a few of the stress factors that can jeopardize the ability of a bird's immune system to fight off disease. Northern goshawks, gyrfalcons, golden eagles, rough-legged hawks, snowy owls, great gray owls, young bald eagles, young red-tailed hawks, and melanistic or albino birds possess an increased susceptibility to this fungus.

Symptoms: Symptoms of aspergillosis vary depending on the severity, location, and stage of the disease. In raptors, they can include a decrease in or loss of appetite, a change in "voice," difficulty breathing (open-mouth, noisy), and weakness or listless behavior.

Diagnosis: Anytime you notice a distinct change in the behavior of a raptor in your care, pay attention! Often, birds don't outwardly

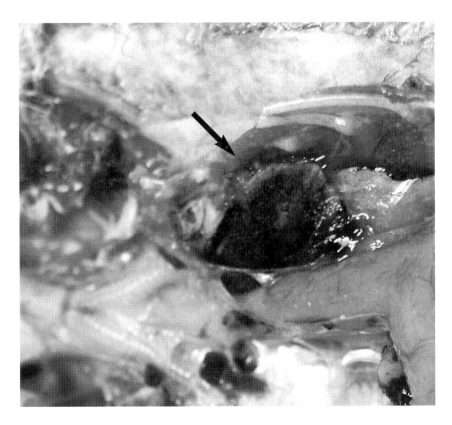

Fig. 7.1
Aspergillus spores growing in the lung of a hawk.

show that they aren't feeling well until they are extremely sick. If you suspect that a bird might have asper or is suffering from some other ailment, don't hesitate to contact your veterinarian. Asper can be diagnosed through a variety of tests. If your veterinarian isn't familiar with diagnosing this disease in birds, contact The Raptor Center for guidance and the most up-to-date information.

Treatment: Currently there is no dependable cure for aspergillosis. Veterinarians are constantly testing new treatments in hope of discovering one that will effectively treat all levels of this disease. Mild cases can be successfully treated with antifungal medications given orally and through a nebulizer. If your bird is diagnosed with asper, your veterinarian can contact The Raptor Center for the most up-to-date treatment information.

Prevention: If the conditions are right, any animal can become infected with aspergillosis. The best way to avoid a bad situation is to prevent it. First, choose a species of raptor that can handle your climate; some species are sensitive to heat, humidity, or other environmental conditions (see chapter 2, Selecting a Bird for Education).

Second, use proper materials for cages and perching surfaces, and keep your bird's housing facilities clean and disinfected. Avoid bedding such as straw, hay, or even wood chips as these all retain moisture and become a breeding field for fungal spores. Also, do not rake or replace the mew floor substrate with the bird present. This activity loosens spores and makes them airborne (chapter 4, Housing). An overwhelming number of spores presented to your bird in a short period of time is often the cause of acute aspergillosis.

Third, reduce your bird's exposure to stressful situations as much as possible.

Fourth, whenever acquiring a new bird in the "sensitive species" list provided above, or moving it to a new location, put it on a three week course of preventative drug therapy. Itraconazole (Sporonox®) administered at 7mg/kg twice a day for five days, followed by once a day for sixteen days will help prevent the disease while the bird adjusts to its new situation. The last thing you can do is to keep an "eagle eye" on your bird to detect potential illnesses early. The sooner asper is diagnosed, the better the chance of a bird's recovery.

Aspergillosis is listed under zoonotic diseases because there is the potential of a person becoming contaminated from a sick bird. However, in order for this to happen, a large number of spores must be inhaled, and people who are immuno-suppressed are at a greater risk. Sick birds generally don't shed spores unless they sneeze, cough, or otherwise discharge respiratory fluids. However, improper protection, such as not wearing a mask when performing necropsies on birds that have died from aspergillosis, or when raking a dirty bird mew, can increase your exposure to asper spores.

Candidiasis

Candidiasis is a yeast infection of the gastrointestinal tract, caused by the organism *Candida albicans*. Most often, this infection is a secondary complication of an injury, illness, or prolonged antibiotic therapy. It is usually not seen in healthy education raptors, but is a common problem in birds that are highly stressed either mentally or due to a physical illness.

Symptoms: Symptoms of a candida infection may include a decrease in appetite, a total loss of appetite, vomiting, difficulty swallowing, food-flicking, and sometimes small whitish plaques in the mouth.

Diagnosis: In raptors, the infection is usually in the mouth and can be diagnosed by swabbing the mouth, culturing it on a special microbiology plate, and incubating it for a few days (figure 7.2). If present, candida colonies will grow on the plate, usually within two days.

Treatment: Fortunately, candida is not difficult to treat. TRC has found that the most effective drug is Nystatin, a liquid given orally at a dose of 1 cc/kg (100,000 units/ml) twice a day for five to seven days. This drug works on contact, so it must touch the affected membranes of the mouth to work. Usually, administering this medicine properly requires restraining the bird; even most "trained" birds won't let you treat them on the glove. Another drug sometimes used is Ketaconazole, a pill given twice per day at a dose of 15mg/kg for seven to fourteen days. TRC has found that this drug is not as effective and prefers to treat candida with Nystatin instead. After the treatment period is over, the infected bird should be recultured to make sure the candida is gone. Your veterinarian will be familiar with this infection and should be able to help you diagnose and treat it.

7.1c Viral Infections

Avian pox

Avian pox is a viral disease caused by a poxvirus. There are several different strains of the virus, and each one seems to be specific for different groups of birds. More prevalent in warm, humid cli-

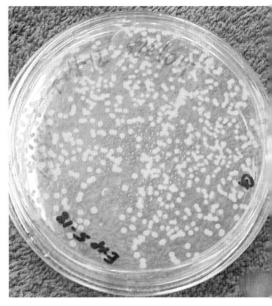

Fig. 7.2
Candida albicans growing on a sab-dex (microbiology) plate.

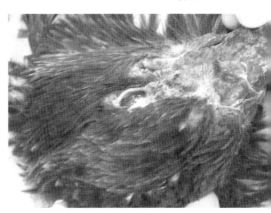

Fig. 7.3 (a)
A young bald eagle suffering from avian pox.

(b)

mates, avian pox can be transmitted by mechanical vectors, such as several species of mosquitoes, or by direct contact between infected and non-infected birds. It can enter the body through broken or abraded skin, or the eye conjunctiva. In a collection of captive birds, it can run rampant and infect the entire group. Thus, birds with pox must be isolated until they have recovered.

Symptoms: A bird infected with avian pox develops wart-like growths on the non-feathered areas of its body such as the feet and face, and sometimes in the mouth (figure 7.3) These growths resemble those of chicken pox in people and can become so large that they prevent a bird from eating, seeing, or using their feet — a death sentence for a raptor in the wild.

Diagnosis: Avian pox can be diagnosed by visualization of the warts. If confirmation is necessary, microscopic examination of a biopsy of the growth and/or virus isolation tests can be performed. However, if pox is suspected, a treatment protocol should be initiated immediately to prevent the virus from spreading and secondary bacterial infection from occurring.

Treatment: There is no known cure for avian pox; it is a virus that must run its course. However, the key to helping your bird through this disease is to prevent further spread and infection, while providing supportive care. TRC recommends treating the warts topically with iodine, which disinfects the areas and dries them out. Wrapping affected feet with an interdigitating bandage is also recommended to keep the feet clean and prevent further spread. Fluids, nutrition (soft foods or tube feeding may be necessary if the mouth is affected), and antibiotics (to prevent the warts from getting a secondary infection) should also be part of the treatment protocol. Contact your veterinarian for assistance.

Prevention: The potential for avian pox can be minimized by reducing the number of biting insects your bird is exposed to. Modifying your bird's outdoor housing to include mosquito netting would help. If a raptor does become infected, separate it from other birds in your collection so biting insects can't transmit the disease to your other avian educators.

West Nile Virus

At the end of 20th century and forging forward into the 21st, perhaps the greatest challenge for raptor caretakers was the arrival of West Nile virus (WNV) in the United States. This deadly virus first attacked raptor populations on the east coast and over a period of

Fig. 7.4

A red-tailed hawk closing one eye due to West Nile virus.

four years, moved its way to the West. Carried by the blood of certain species of mosquitoes, WNV has been the cause of mortality for many raptors, both in wild populations and captive collections. As of the writing of this book, there are still many unanswered questions about the modes of virus transmission.

Symptoms: Birds that have contracted the disease can show a variety of symptoms depending on the species and the length of infection. These symptoms may include depression, sleepiness, lack of appetite with corresponding weight loss, pinched off blood feathers, an elevated white blood cell count, an enlarged spleen, green urates (indicating liver necrosis), weakness in the legs, a loss of use of one or both wings (due to the virus attacking the brachial plexus nerve bundle in the shoulder area), exaggerated aggression, head tremors, seizures, keeping one eye closed due to developing blindness (retinal degeneration has been seen in hawk species) (figure 7.4) and death.

A bird infected with the virus is not going to exhibit all of these symptoms; some birds may present with only pinched blood feathers and an elevated white count, while others may quickly develop neurological signs and die.

Diagnosis: There is currently no reliable WNV diagnosis for a live bird. To confirm the disease, special tests must be conducted on the brain and/or various other tissues. Many people test their birds for the presence of WNV antibodies in the blood. If antibodies are present, it does not mean the birds have or had the disease. It merely means that the bird has been exposed to it and its immune system responded by producing antibodies.

Treatment: If you suspect WNV in your avian educator, take it to your veterinarian immediately. Proper supportive care in the early stages of infection can help your bird survive this disease. For the most current treatment protocol, contact The Raptor Center.

Prevention: To help keep your bird safe from WNV, there are a few things you can do. First, if feasible, cover its mew with mosquito netting to reduce the number of potential carrier mosquitoes your bird will be exposed to. The Mosquito Magnet® (American Biophysics Corp.) can also be purchased and placed near the mew to reduce the number of mosquitoes in the immediate area. These units attract female mosquitoes by emitting carbon dioxide, moisture, heat, and an attractant over a short range. The mosquitoes are then trapped. These units can be purchased from a variety of sources listed on the web.

Second, do not feed your raptors English Sparrows or European starlings. These non-native species are carriers of WNV and it is not yet known whether your raptor could contract the virus from eating an infected bird.

A third step you can take to prevent your raptor from contracting WNV is to vaccinate it. Currently, birds of prey are being vaccinated with the Fort Dodge® equine vaccine. If a bird has not been previously vaccnated, two doses, three to four weeks apart should be given. Then, in subsequent years, a bird can get a single booster shot. The vaccine should be administered intramuscularly (the pectoral muscles are common injection sites) a few weeks before the virus has typically appeared in your area. In the upper Midwest, for example, vaccinations are recommended in early June. Suggested vaccine volumes are as follows and are based on a bird's size: <200g: 0.3cc (0.010 oz.); 200–400g range: 0.5cc (0.016 oz.); 400–650g range: 1.0cc (0.033 oz.) divided between two injection sites; >650g: 1.0cc (0.033 oz.).

Young birds, four weeks of age and older, can also be vaccinated and require two injections, three to four weeks apart.

7.1d Special Considerations

There are few diseases that should be mentioned, not because they commonly occur in raptors, but because they are high profile diseases that people often have questions about.

Newcastle Disease

Newcastle disease is caused by a virus that enters a host through either its respiratory or its gastrointestinal tract. The most common carriers are free-ranging birds, especially pelicans and cormorants. However, free-ranging waterfowl, parrots, songbirds, and owls have been identified as potential carriers of the virus (they don't contract the disease but can spread the virus by shedding it in their mutes).

The virus can also be spread by the wind, insects, and humans. There are no known cases directly linking the disease in humans to a raptor carrier. Humans have, however, contracted the disease from other carriers and developed a characteristic severe eye infection. If outbreaks of Newcastle disease do occur in your area (most likely in other species), it may affect your ability to travel with your raptors if you conduct off-site programs.

Psittacosis

Also known as "Ornithosis" or "Chlamydophylosis," this bacterial infection can be transmitted through a bird's mutes, nasal discharges, or tissues. It occurs most commonly in psittacines, but can affect raptors. The disease can cause sleepiness, a reduction in appetite, respiratory problems, watery green droppings, or diarrhea tinged with blood. Humans can contract the disease from birds if they inhale bacteria that have been aerosolized. In people, the infection often settles in the lungs and presents flu-like symptoms, although eye infections have also been reported. Psittacosis is often successfully treated with tetracycline antibiotics.

Rabies

When people think of wildlife diseases, rabies always comes to mind. Movies have dramatized this disease and created a very misleading picture. Infected animals are projected as raging mad, foaming at the mouth, and totally out of control. Animals can carry the rabies virus without showing these pronounced signs. They themselves might not get sick with the virus, but are quite capable of spreading it through their saliva. Raptors are not immune; although they can carry the virus, there are no known reports of raptors becoming ill or transmitting the virus to humans. Humans have become infected by other animals such as dogs and bats, and a life-threatening viral encephalitis results.

7.1e Zoonotic diseases

Avian TB, salmonellosis, tetanus, aspergillosis, West Nile virus, Newcastle disease, psittacosis, and rabies are classified as zoonotic, meaning that there is the potential for the disease to be transmitted from one animal to another. This may sound frightening, but the occurrence of transmission from raptors to humans is extremely rare. Despite this, it is important to understand how transmission of zoonotic diseases generally occurs and what steps can be taken to prevent it.

Transmission

Zoonotic diseases can be transmitted from raptors to humans either directly or indirectly. Direct transmission usually occurs through bites, scratches, or punctures (these provide entry points for disease organisms), or from handling tissues of infected birds. Indirect transmission usually occurs through contact with contaminated mutes, contaminated ground substrate, or open sores on a bird.

Proper handling methods (see chapter 6, Maintenance Care), sanitation (chapter 4, Housing), and personal hygiene are definitely the keys to prevention. If you develop an illness, make sure to let your doctor know that you work with birds. This piece of information may help diagnose the problem.

Prevention

Zoonotic diseases might sound scary, but there are steps you can take to decrease the possibility of encountering them:

- Make sure any bird you receive has had a complete physical exam and appears healthy. Find out about its background, medical record, and handling history. Don't accept a bird unless it has a clean bill of health.

- Whenever you handle a raptor, wear protective clothing. Leather gloves, gauntlets, and welder's jackets (with larger, untrained display birds) will help protect your body from potential punctures and scratches that could be entry points for disease organisms.

- Do not let your bird's mew get extremely dirty with a build-up of feces. Many organisms are shed in feces and can be aerosolized during the cleaning process.

- Always wash your hands after handling a bird, cleaning its mew, handling or preparing its food, etc.

- If you house your bird indoors, be sure that its enclosure is sufficiently ventilated (see chapter 4, Housing) and that the surrounding room is also well ventilated. Proper ventilation helps minimize the presence of airborne disease organisms that either you or your bird could inhale.

- Wear a mask when raking mews, replacing gravel, etc.

- Make sure that you and everyone else who handles your bird has an up-to-date tetanus immunization and does not have a compromised immune system.

7.2 MANAGEMENT-RELATED INJURIES AND ILLNESSES

Many medical problems faced by captive raptors result directly from management issues. It can be challenging for a raptor caretaker to meet all the physical and mental needs of a raptor as it faces different weather conditions throughout the year and goes through its juvenile, breeding, and geriatric stages of life. What may have worked for a long time will suddenly change and a caretaker must have strong observation skills to catch when changes in management are needed.

Presented here are the most common management-related medical problems (in alphabetical order) that a captive raptor may face.

7.2a Bumblefoot

Probably the most sensitive part of a captive raptor is its feet. In the wild, raptors maintain healthy feet by utilizing perches of different diameters and surfaces, thus preventing pressure sores from developing. Environmental factors, such as sunshine and moisture, also help to keep the skin on a raptor's feet healthy. In captivity, however, raptors often have limited perching options and less exposure to the weather, increasing the probability that they will develop foot problems.

Bumblefoot is a general term for any degenerative or inflammatory condition of a bird's foot. It usually begins as a mild redness, thickening of the skin (excess keratinization), abrasions under the talons, or swelling of the metatarsal pad (bottom of the foot) (figure 7.5). A pink coloration to the bottom of the foot indicates thinning of the skin leading to blistering and an open wound just waiting for infection. In addition, overgrown talons can cause self-inflicted puncture wounds on the toes or bottoms of the feet, which can easily become infected and cause surrounding tissue to become diseased and die. If these conditions go undetected and untreated, infection can extend into the bones of the foot and cause life-threatening bone infections.

It's critical, therefore, to maintain constant vigilance over the condition of your bird's feet. Both you and your bird will benefit if you to learn how to recognize and treat the early signs of bumblefoot. More extensive foot problems will require consultation with and treatment by a qualified veterinarian. Keep in mind that many foot problems are difficult to cure and require time and patience. In the case of bumblefoot, an ounce of prevention is certainly worth a pound of cure.

Causes
There are many management factors that can lead to bumblefoot if not properly addressed and monitored. Listed below are the most common causes:

1. Improper perches. Perch size, shape, and covering all can influence the bird's weight distribution on its toes and metatarsal pads and the subsequent amount of wearing on the skin (refer to chapter 4, Housing, for specific recommendations).

2. Inadequate housing. Bruising on the bottoms of the feet can

Fig. 7.5
Common forms of bumblefoot in raptors.
(Illustration by Gail Buhl, photos by author)

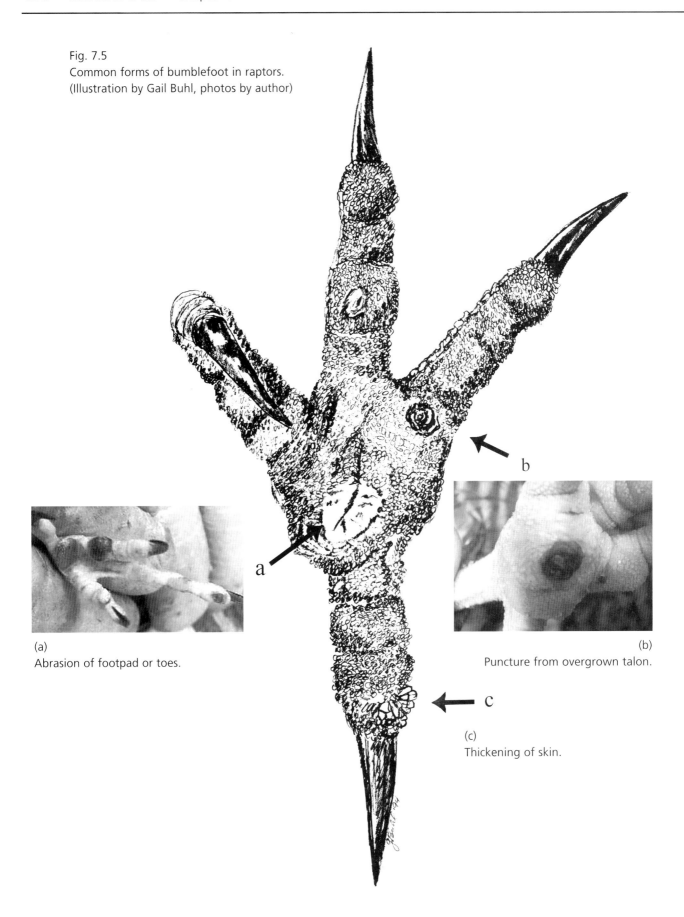

(a)
Abrasion of footpad or toes.

(b)
Puncture from overgrown talon.

(c)
Thickening of skin.

lead to bumblefoot. This can occur if a free-lofted bird is housed in an enclosure that doesn't allow flight, so that the bird merely bounces from perch to perch. It will also occur if a tethered bird persistently bates from a perch onto a hard surface. In addition, foot damage can result if a raptor jumps against a wire enclosure and hangs from the wire, causing skin trauma (as evidenced by abrasions on the toes under the talons) and potential rupture of the flexor tendons in the toes. This can be prevented through proper cage design (chapter 4, Housing).

3. Overgrown talons. Talons that are too long or too sharp can cause improper weight distribution on the bottom of the feet (especially in falcons) or result in self-inflicted puncture wounds to toes or footpads.

4. Obesity. Raptors held in captivity can easily become overweight if not managed properly. Inactivity, accompanied by an ample food source, causes an unnatural weight gain that puts excessive pressure on a raptor's feet. This pressure leads to thin, worn skin that can evolve into open sores.

5. Uneven weight bearing on the feet. Wing amputations or nerve damage, an injured pelvis, a leg or toe injury, or arthritis can all result in greater weight bearing on one foot. If your bird develops a pelvic, leg, or toe injury you should protect the other foot by applying a bandage or protective "shoe" during healing. Contact your veterinarian or an experienced rehabilitator for assistance.

6. Poor living conditions. Raptors maintained with an inadequate diet lacking essential vitamins and moisture, limited access to sunlight, and a dirty living environment are at high risk for developing foot problems. Good nutrition, water, sunlight and a clean living area (which minimizes the presence of bacteria that could invade a small cut or puncture) are critical in keeping your bird and its feet healthy.

Symptoms

Regular checks of your bird's feet will reveal the more obvious symptoms of foot problems, such as punctures, corns, and pink, thin pads on the bottoms of the feet. However, your bird might also provide other signs. If it appears to favor a foot (constantly holding one up), lies down, leans to one side, or refuses to hold onto food, it might be telling you that it has sore feet or a leg injury. Be aware of even slight changes in your bird's behavior; often, they indicate some kind of problem.

Treatment

There are several techniques for treating early signs of bumblefoot. Most of the products required are available at your local pet store or drugstore. Pink, thin pads can be treated topically with the daily application of Tuf-foot®, which should help toughen thin skin. This can be applied by one or two people. If a helper is available, hold the bird on your gloved hand and slowly and gently lift up one foot. Your helper can dip a cotton-tipped applicator into the bottle and apply it to the pink areas of the foot. Repeat this procedure with the other foot if needed.

If the bird is too nervous with the helper near, you might want to try to treat it yourself. You'll have to apply the medication by slipping the applicator under the bird's feet as it perches on your fist. As you try to do this, the bird will often lift its foot up for a brief time to get away from the irritation of the moving applicator under its foot. That's your chance to apply the medication.

Dry, cracking feet should be moisturized with the topical application of a moisturizer such as bag balm, Protecta-Pad®, A and D ointment, a calendula cream mixture (calendula, echinacea, hypericum cream), or Aloe Vera Gelly®. Natural emu oil can also be used and has proven to be a good moisturizer as it penetrates well into the skin, is not irritating, and has anti-inflammatory properties (appendix E). These products can be either applied with a cotton-tipped applicator, or rubbed into the feet with your hands (if your bird has a good temperament). Again, this treatment can be done by one or two people. However, during cold temperatures, do not immediately return the bird outdoors following treatment. Let the moisturizer completely absorb into the skin first to prevent frostbite.

If the pinkness or dryness persists, or if the bird's feet develop blisters (thin, white areas on the pads), scabs, or other wounds, consult a veterinarian as soon as possible. If your veterinarian isn't familiar with the techniques for treating foot problems, contact TRC for assistance.

Prevention

Preventing bumblefoot in raptors under your care is well worth making every possible effort. Choosing the proper perches for a particular species is very important, as is creating a healthy environment in which to house the bird. Housing should be designed to minimize the potential for self-destructive behavior. Vertical bars or solid walls will curb wire-hanging tendencies. For tethered birds that have a tendency to bate, indoor/outdoor carpeting or short artificial turf can be placed around the perch to reduce bruising of the feet. Strict sanitation will minimize the potential for bacterial infections in enclosures (see chapter 4, Housing). Finally, a well-balanced diet with adequate vitamins A and E, together with fresh water for bathing, will help ensure that the epithelial tissue of your bird's feet stays healthy. To make sure that your management plan is meeting your bird's needs, foot and health checks must be conducted regularly (see chapter 6, Maintenance Care).

7.2b Cold Weather Anemia

Housing a raptor outside during cold winter months warrants special attention due to a condition called cold weather anemia. Anemia is the lack of a sufficient number of red blood cells. It can be caused by an illness, a lack of proper nutrition (an inadequate amount of food, or poor quality food such as only breast meat, legs, head, and neck of prey), exposure to a sudden temperature change, or a major break in the bird's routine. Anemia is most likely to be seen in falconry birds and demonstration raptors from mid-November through March in cold climates, when they are maintained at their "flying weights" for a prolonged time while housed outdoors in an energy-demanding climate. If not treated immediately, this anemia can be fatal.

Symptoms
Various symptoms can indicate anemia in a raptor, including change in behavior, lack of appetite, unwillingness to fly, weakness, "almond-eyed" appearance, minor tremors or shaking, sudden and noticeable prominence of its keel (breastbone), or excessive sleeping during the day. Your veterinarian can quickly diagnose anemia by evaluating a blood sample for the percent of red blood cells and total solids present. In a raptor, red blood cells should range from 30–48 percent (large falcons tend to be on the higher range). Birds diagnosed with anemia, have less than 30 percent red blood cells. The normal range for total solids is between 3.0 and 4.5 g/dl.

Treatment
As with all medical conditions, anemia should be treated by a veterinarian. The general treatment protocol includes the administration of fluids either subcutaneously or intravenously, iron dextran (10mg/kg intramuscularly once), vitamin B (10mg/kg subcutaneously once per day for a few days), warmth, and small amounts of easily digestible food (slurry tubed into the stomach, clean meat — no casting material). Whole food should not be given until the bird has been well hydrated and is metabolizing the easily

digestible food. This may take a few days. In severe cases where the percent of red blood cells is less than 20 percent and the total solids are less than 1.0 g/dl, a blood transfusion and plasma expander (Hetastarch®) may be needed to save your bird. Three weeks is an average recovery time.

Prevention

To prevent cold weather anemia from developing in your avian educator, feed it a high-quality, varied diet (whole prey) supplemented with vitamins such as Vionate or Vitahawk (appendix B), keep good consumption records, and monitor its health and behavior daily. If your bird is flown for demonstrations, oscillate its weight regularly (don't constantly keep it at a flying weight; fatten it up once a week), give it time off, maintain it at a higher flying weight during the winter than at other times, and provide it with sufficient warmth.

7.2c Frostbite

In northern latitudes, where temperatures often fall below freezing and wind chills dip below zero, captive raptors are vulnerable to physical ailments that result from prolonged exposure. Exposed skin is highly susceptible to frostbite during times of excessively cold temperatures and low wind chills. Species with bare legs, such as American kestrels, bald eagles, broad-winged hawks, common barn owls, peregrine falcons, osprey, red-tailed hawks, and vulture species are more prone to developing frostbite on their feet than are species with feathered legs and feet (figure 7.6a).

A free-lofted raptor in captivity is most likely to develop frostbite from ice buildup under its anklets or on its perches. A tethered bird is at greater risk, and can get frostbite not only from these causes, but also from restricted movement on a cold, moist surface if its leash or jesses have become tangled, or get wet and freeze. A tethered bird that gets twisted can also get frostbitten wing tips and/or feet if it bates or struggles constantly on a snowy or icy surface. In addition, Harris's hawks and some falcon species are especially at risk for developing frostbite on their wing tips. In cold climates, species sensitive to cold temperatures must be provided with a heat source to prevent frostbite (chapter 2, Selecting a Bird for Education).

Symptoms

You'll usually start to see signs of frostbite a few days after the exposure. It appears as either orange/pink discoloration of the toes, or edema — white, fluid-filled skin that is cool to touch — on the tips of the wings (over the metacarpal bones) (figure 7.6b).

Fig. 7.6
Frostbite.

(a)
Two severely frostbitten toes of a bald eagle.

(b)
Wing tip edema in a peregrine falcon.

Treatment

If you suspect frostbite, bring the bird inside immediately and contact your veterinarian. If the bird's feet are of concern, soak them in warm (not hot) water for thirty to sixty minutes. If you're worried about the bird's wing tips, apply a warm compress to the area for fifteen minutes. A microwaveable warm pack can be used, or more simply, a wet towel heated in the microwave until it is warm to the touch, but not burning.

Once frostbite occurs, you can do little to correct the damage, but you can prevent additional tissue or circulatory compromise. In mild cases caught early, tissue can survive and the discolorations disappear. In moderate cases, a patch of skin may die and tendon and muscle function may be affected. Skin can grow back; in these cases, the tendon and muscle function is of greatest concern. In severe cases an entire appendage is frozen, leading to a loss of circulation and appendage death (figure 7.6a). It often turns black and dies within two to three weeks of the frostbite incidence and either sloughs off or should be removed surgically. If a raptor loses a foot to frostbite, it should be humanely euthanized.

TRC treats frostbite patients with the following protocol:

- Moist warm packs are applied at least twice daily to the affected area for fifteen minutes at a time and Aloe Vera Gelly (appendix E) is applied.

- The bird is put on several medications including:

  - Antibiotics to prevent secondary infection from loss of a healthy skin barrier. (clavamox 50–75mg/kg twice a day)

  - A vaso-dilator to improve circulation in the appendages (isoxuprene 5–10mg/kg once per day)

  - An aspergillosis preventative if it is a sensitive species (itraconazole at 7mg/kg twice per day for five days followed by once per day for sixteen days), and

  - A non-steroidal anti-inflammatory to help reduce tissue swelling

- In addition, DMSO (dimethylsulfoxide) is applied topically with a cotton-tipped applicator to the affected area to improve circulation and reduce swelling. If you are going to apply DMSO, it is strongly advised to wear gloves to prevent absorption through your skin (some people have reactions to it such as developing a garlic taste in their mouth or getting severe headaches). DMSO should only be applied to the affected area once per day for no more than three days to avoid tissue sloughing.

Prevention

There are a few precautions you can take to minimize the potential for frostbite. To protect your bird's toes, keep perches ice-free by either promptly removing ice manually or providing a heat source to prevent ice from forming (chapter 4, Housing). Provide your bird with an adequate shelter box constructed and installed to minimize ice buildup on at least one perch. If your bird is tethered, check it often to make sure it isn't tangled and to remove ice or snow from the area to which the bird has access. If a bird has foot bandages due to sore feet or thin skin, don't house it outside. Ice buildup under the bandages can easily lead to frostbite. Also, if you need to apply any creams or ointments to your bird's feet bring the bird indoors to treat it and let the cream or ointment completely absorb before returning the bird outside. Finally, as mentioned in chapter 4, Housing, do not offer your bird a bath outside during freezing temperatures.

7.2d Heat Stroke

High temperatures and humidity can cause great physical distress to a raptor and become life threatening. Heat stroke is a disturbance in an animal's ability to cool itself and can result in a dangerously elevated body temperature followed by sudden death. Believe it or not, even in Midwestern states, summer temperatures and humidity can cause heat stroke if a raptor isn't properly cared for.

Symptoms

Excessive panting together with droopy wings, weakness, and collapse are usually signs that heat stroke has occurred. Keep in mind, however, that mild panting and gular fluttering are normal methods of heat dissipation for hawks and owls, respectively. Also, certain species, such as burrowing owls, droop their wings over their legs to shade them. This is a normal behavior pattern. Be aware of these species behaviors so you can recognize whether or not your bird is in distress.

Treatment

If you identify heat stroke in your bird, you must act quickly. There are several steps you should take immediately:

1. Contact your veterinarian by phone for assistance.

2. Either spray your bird's feathers with cool water (drench them) or hold the bird under a faucet or shower. It's critical to wet the down feathers close to your bird's skin to facilitate heat loss.

3. Immerse your bird's feet in cool water. Birds can lose heat through their feet.

With assistance from your veterinarian, you can take other measures to reduce the bird's body temperature. In addition, your veterinarian will provide medical support to prevent shock when cooling your bird. The shock related to drastic changes in body temperature can also be fatal and must be treated.

Prevention

No matter what your climate, you can help prevent heat stroke by taking the following steps:

*   Provide your raptor with a shelter box in its enclosure, so it will have shade. The best orientation for a shelter box is facing the morning sun but avoiding the hot sun later in the day. If your bird is tethered, make sure it has access to shade throughout the entire day.

*   Give your bird access to water for drinking and bathing during warm temperatures (chapter 3, Diet).

*   Never leave your raptor in an unattended car on hot and/or humid days, even in the shade. The increased temperature and reduced ventilation in a parked car can rapidly lead to overheating and death. We strongly recommend that on warm days (temperatures over 75°F/23.9°C), you transport your bird in an air-conditioned vehicle.

*   If flying your raptor is a part of your education program, give the bird a rest on excessively hot and humid days. The heat generated from exercise in such weather can lead to heat stroke.

*   Northern raptor species (gyrfalcons, rough-legged hawks, snowy owls, great gray owls, boreal owls) are less tolerant of heat and can overheat even in relatively cool temperatures (70°F/21.1°C and above), depending on relative humidity,

Fig. 7.7
An adult bald eagle exhibiting clinical signs of severe lead poisoning.

wind, and cloud cover. If you live in a warm climate, these species are not the best choices for education birds. Some species don't tolerate weather extremes (chapter 2, Selecting a Bird for Education).

Maintaining a healthy raptor in captivity is a major responsibility. Keep in mind that temperatures and climatic conditions that feel comfortable to you might not feel comfortable to your bird, and might even cause life-threatening illnesses. Use common sense. Be especially aware of quick changes in temperature. Educate yourself about which species can best handle your climate and then monitor your bird closely during extreme weather conditions. Following these suggestions will help protect your bird from serious weather-related conditions.

7.2e Poisoning

Aerosol and gaseous toxins

Carbon Monoxide: Occasionally, the news reports an incident of carbon monoxide poisoning (CO) in people. Faulty furnaces that emit this odorless gas are one major cause. Now most households have one or more CO detectors. Did you know that your raptor is susceptible to CO poisoning as well?

Poor ventilation during vehicle transport can result in death to your bird. This has most often been reported in Harris's hawks that traveled in covered beds of pick-up trucks. Exhaust fumes that were not vented properly built up in the beds and caused fatal CO poisoning. In raptors, there are no symptoms reported as most victims are discovered after they have died.

To prevent this tragic death from occurring in your bird, make sure that your vehicle's exhaust is properly vented and stays clear of your bird at all times. Do not keep your vehicle running while loading or unloading your avian passengers, and do not place your bird (in or out of its crate) near the exhaust pipe.

Other Substances: Birds have very refined and sensitive respiratory systems and can also be affected by a variety of other fumes. Fumes from disinfectants (such as concentrated bleach), heated Teflon-coated pans or light bulb covers, moth balls, oil-based paints and varnishes, and exhaust fumes from vehicles or machinery can all create illness in your raptor, and in extreme cases, death. Be conscious of your bird's location in relation to activities that may result in the presence of such fumes. Remove your bird when disinfecting its mew, use water-based products for protecting wooden perches and travel boxes, and keep your bird away from heated Teflon, and any type of exhaust fumes.

Lead

Most people think of lead poisoning as a problem raptors face only in the wild. However, there are also a few ways that captive raptors can be exposed to this highly toxic metal.

In chapter 3, Diet, it was not recommended to feed your bird wild game birds or venison. In both cases, it is risky because the meat could contain lead. Lead is a soft metal and when a game animal is shot, lead is deposited over the range of tissue from the point of impact until the shot is lodged. A raptor does not need to ingest a lead pellet to get sick. If it eats meat that has lead residue/fragments, it could ingest enough to cause illness. Captive golden and bald eagles, peregrine falcons, and several vulture species have been treated at The Raptor Center for lead poisoning that resulted from being fed these types of food items.

Lead is also present in other forms, including lead-based paint. Even though modern paint products are lead-free, many older buildings still have lead-based paint on the walls. Some species of raptors, such as vultures and caracaras get easily bored and start picking at things in their enclosure. One attractive item is paint chips that have peeled off the walls. Therefore, it is not recommended to house birds, even short term, in older buildings until old paint has been removed (and the building has been renovated to reduce moisture and mold and provide a good ventilation system).

Symptoms: Raptors that ingest lead can show a variety of clinical signs including green mutes, general depression, respiratory distress (long inspirations and forced expirations, open mouth breathing), vocalizations with each breath, droopy wings, anemia, an almond-eyed appearance, and blindness (figure 7.7). Lead also causes some birds to ingest inedible items (foam, leather, rope from perches, etc).

Diagnosis: Lead poisoning can be diagnosed by taking a blood sample and having a laboratory test it for the presence of lead. If you suspect lead poisoning, contact your veterinarian immediately for assistance.

Treatment: There are several variables that determine how severe a lead poisoning incident will be. These include the amount of lead ingested, the length of time the lead stays in the gastrointestinal tract (raptors can remove the lead in a pellet if the

GI tract remains functioning), and the amount of time that lapsed before diagnosis and treatment. As mentioned above, even a small amount of lead can cause fatal lead poisoning. It has been TRC's experience that if a bird's blood lead level is greater than 1ppm (part per million) and the bird exhibits typical signs of lead poisoning, the prognosis is extremely poor and most birds don't survive; the lead has already caused severe and irreversible damage to vital tissues and organs.

The basic treatment for lead poisoning includes the administration of a chelating agent (100mg/kg Ca EDTA diluted in saline, given subcutaneously or intravenously twice daily for five consecutive days before rechecking the blood lead level), itraconazole, an aspergillosis preventative (given 7mg/kg twice per day for 5 days, followed by once per day until all the lead is removed), and daily treatments of vitamin B, C, zinc, and a non-steroidal anti-inflammatory.

7.2f Sour Crop

A crop is an outpocketing of the esophagus in hawks (and many other species of birds, but not owls) that acts as a storage bin for food before it is digested. You may hear the expression, the hawk has a "full crop," or the bird "gorged." Both of these terms refer to a large meal that a hawk has eaten and that visually expands the crop. Another term you may hear, is "putting a crop over." To help the food move from the crop to the stomach where digestion begins, hawks mechanically push it down by moving their necks back and forth and up and down.

Cause

There are times when a bird's digestive system shuts down and as a result, food sits in the crop for twenty-four hours or more. When this happens, bacteria begin to flourish and the food develops a sour smell. Crops can become sour for several reasons: a bird is sick or anemic (it has a low number of red blood cells), a bird is very thin and its system is overwhelmed by a large meal that it can't metabolize, an obstruction is present (such as a large bone, gravel, etc) preventing the mechanical movement of pushing crop contents down, or the crop ruptured (this can happen from the ingestion of a large bone, or collision with a wall, etc. after eating a full crop). A sour crop in a captive raptor is usually the result of poor management practices.

Treatment

A sour crop must be treated or your bird could die from a severe

Fig. 7.8
A syringe case cover designed to protect a damaged talon.

(a)
Syringe case cover.

bacterial infection. Your veterinarian should be able to assist you in this process. The contents of the crop need to be removed, the crop flushed with saline, and the bird given fluids to prevent dehydration, antibiotics to treat the bacterial infection (metronidazole at 50mg/kg twice per day for a minimum of five days), and easily digestible nutrition for a few days (slurry, small meals of meat without bones or fur).

To help the gastrointestinal system get back on track, a motility aid such as metaclopromide (2mg/kg intramuscular), ranitidine (0.2–0.5mg/kg subcutaneous, intramuscular or intravenous), or an oral crop drench of fennel tea (steep one teaspoon of crushed tea in one cup boiling water, strain to remove grains, cool to body temperature, and tube into the crop at a volume of 1–3 percent of the bird's body weight) can also be administered.

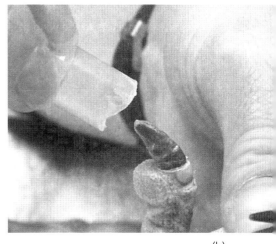

(b)
Fitting the case to fit comfortably over the talon.

7.2g Talon Sheath Loss

As mentioned in chapter 6, Maintenance Care, talons have an outer sheath (consisting of a protein called keratin) that protects the talon bone. It is similar to a fingernail and can be lost or damaged.

Cause
A captive raptor can lose a talon sheath if it grabs tight to a perch or blanket/towel when being grabbed or if a caretaker tries to pry talons off of a glove, towel, or perch too harshly. A lost talon sheath appears painful and results in a fair amount of bleeding.

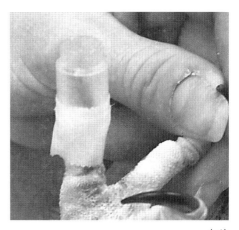

(c1)
Securing the case with tape and vet wrap, leaving the end open.

Treatment
Your mission is to stop the bleeding, keep the wound clean, and when it is dry, coat it to prevent bacteria from entering. Infections at the nail bed are common if the area is not properly cleaned and protected while it heals. If you do not have the resources to adequately treat this injury, try to control the bleeding by applying pressure to the area for several minutes and contact your veterinarian. TRC recommends the following protocol to treat talon damage from a lost sheath:

1. Stop the bleeding by applying pressure, Kwik Stop antiseptic powder, or tissue glue to the area. The powder and glue will only work if you have at least a few seconds where the site is free of fresh blood.

2. Gently clean the affected toe and talon by dabbing either a dilute iodine solution or dilute Nolvasan on it with a cotton-tipped applicator.

(c2)

Fig. 7.9
Applying wrist protectors to the wing of an osprey. (Gail Buhl)

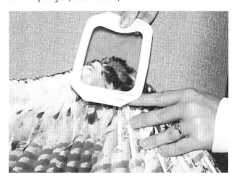

(a)
With wing extended, place one piece of Tegaderm over wrist.

(b)
Place one piece of Microfoam tape directly over the top of the wrist.

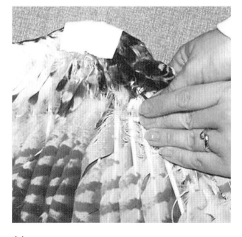

(c)
Place one piece of Microfoam tape on each side of wrist, overlapping the first piece applied.

3. Cover the toe and talon to prevent re-injury as it heals. TRC recommends applying a small section of plastic syringe case over the talon (filing the case down so there are no sharp edges), taping the bottom edge to the toe and covering it with vet wrap, leaving the top end open so air can reach it (figure 7.8).

4. Once the talon bone is dry, apply a few layers of clear fingernail polish to strengthen and protect it.

You can expect the nail bed to be oozy for a few days. However, if the toe tip gets swollen, is reddish in color, and/or is warmer than the other toes, infection may be setting in and antibiotics may be called for. Contact your veterinarian for assistance. A talon sheath will grow back, although it takes a long time (up to a year) and depending on the damage to the underlying talon bone, may only grow back to part of its original length.

7.2h Wrist Injuries

When confronted with a situation perceived as potentially dangerous, raptors have one of two responses: fight or fly. Many species choose to fly to remove themselves from imminent danger. In captivity, this same behavior occurs, with the stimulus often being people, other birds, a gloved hand, a noisy lawn mower, etc. When the flight response occurs in a limited space, a bird can hit its wings on the enclosure walls, ground, roof, or furniture (perches, shelter box) and injure its wrists in its panic to find safety (figure 4.14). This more commonly occurs in newly acquired birds that are still adapting to their surroundings, especially if they are free-lofted. Birds have thin skin and it does not take much trauma to open the skin over the wrist joints and allow infection to set in.

Treatment
If caught early, wrist wounds can be cleaned and sutured closed by your veterinarian. If they are not found right away, swelling and infection can set in and must be resolved before skin closure can occur. Contact your veterinarian immediately for a treatment plan. If untreated, wrist wounds can lead to osteomyelitis of the carpal (wrist bones) and permanently reduce wing extension.

Prevention
If you notice that your bird is flighty in its new situation, protective wraps can be temporarily applied to its wrists while it is adapting and/or being trained. Supplies needed include Microfoam tape®,

Durapore tape®, and Tegaderm®, all 3M Industries products. To apply wrist protectors, follow these easy steps:

1. Fully extend the wing and apply one piece of Tegaderm lengthwise directly over the wrist bones (figure 7.9a).

2. Place three pieces of Microfoam tape lengthwise over the Tegarderm, the first directly over the wrist bones and the other two overlapping each edge of the first piece. Do not make the Microfoam longer than the edge of the Tegaderm (figure 7.9b,c).

3. Place another piece of Tegaderm lengthwise over the Microfoam to hold it in place (figure 7.9d).

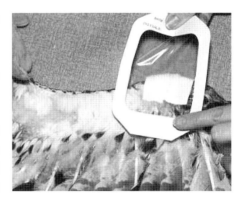

(d)
Cover microfoam tape with a second piece of Tegaderm.

Tegaderm works well for these protectors because it adheres well to the feathers and maintains its integrity as the bird flaps it wings, bathes, etc. The negative side to Tegaderm is its cost. A second, but less secure option for wrist protectors is to place the Microfoam tape directly on the feathers over the wrist and secure it down by placing a border of Durapore tape on all edges of the Microfoam. The Durapore tape does not stick as well, is easier for the birds to pick off, and tends not to stay on long if the bird bathes.

Wrist protectors can remain on your bird as long as you feel is necessary. Fully flighted raptors can even wear these protectors in their free-loft mews; the protectors do not interfere with their flight ability. Wrist protectors are most commonly used for accipiters, eagles, osprey, and vultures, but can be applied to any hawk species. Many owl species, on the other hand, tend to be less accepting of bandages in general and often pull bandages off within a few minutes. Therefore, it is best to avoid using protective wraps. If an owl species injures its wrist, it may need more extensive bandaging of the area until it heals (bandaging tape and or vet wrap covered with a layer or two of duct tape often does the trick; do not put duct tape directly on the feathers).

7.3 NUTRITIONAL DISORDERS

Remember when you were young and your parents always told you to eat your vegetables? Well, as most of us realize now, our parents were trying to provide us with the healthy, balanced diet we needed for our growth and development. Even now that we are out of that growth stage, we still need a balanced diet to keep our bodies and minds fit and meet our changing nutritional needs as we progress through the various stages of our lives.

That same concept can also help you provide the best possible care for your bird. Just like mammals, raptors require a well-balanced diet to maintain their health (chapter 3, Diet). Like those of humans, birds' nutritional needs also change during different stages of life, including the annual molting season, weather extremes, and aging. Either a deficiency or an excess of specific nutrients can cause health problems, some of which can prove fatal with time.

7.3a Nutritional Deficiencies

We've already mentioned that your bird needs certain basic nutrients to keep it healthy. If these essential nutrients are lacking, the bird's body won't function properly. The bird might become sick or exhibit poor skin color, feather condition, or beak and talon condition. Since a raptor's meat diet usually provides sufficient protein, carbohydrates, and fat, most deficiency disorders originate from a lack of vitamins or minerals.

Vitamin D
Vitamin D is essential for absorbing and using calcium and for developing and maintaining healthy bones. A diet consistently deficient in this vitamin leads to weakened bones. In time, vitamin D deficiency can also lead to convulsions caused by low blood calcium. Diets consisting of fish and bird prey items tend to have higher levels of usable vitamin D than all-mammal diets do. Therefore, if your bird is on a strictly mammal diet and/or has limited exposure to sunlight (which metabolizes a precursor to vitamin D), a supplement may be recommended (see chapter 3, Diet). Revising your bird's management plan to include some bird prey items and a general vitamin supplement, as well as providing sufficient sunlight, is the best way to provide sufficient Vitamin D.

Vitamin A
Vitamin A deficiencies can also occur. Regularly feeding your bird a diet of meat (without organ meat such as liver) can lead to poor-quality skin, pale skin color, corns on the feet, and lesions in the mouth and crop. A whole prey diet is critical to keeping your bird healthy. Recommended diet supplements include animal liver, day-old chicks, and a vitamin supplement, such as Vitahawk Maintenance® (appendix B).

Vitamin B
Vitamin B deficiencies can occur as well. The most common vitamin-B deficiency is a shortage of dietary B1 (thiamine), which can lead to nervous-system disorders. Some raptor diets, such as all fish

diets, can be deficient in this vitamin. Different species of fish have different vitamin B contents and often the amount is low (it is based on the amount of thiaminase a fish species has; this enzyme breaks down thiamine). Freezing the fish then degrades any vitamin B that is present, making it unusable. Therefore, raptors kept strictly on a frozen fish diet (such as osprey) require supplemental thiamine once per week (10mg thiamine/kg bird). Also, other raptor species kept on a diet of day-old chicks or muscle meat may develop a thiamine deficiency. A varied diet is the key to keeping your raptor healthy.

Minerals

Minerals, particularly calcium and phosphorus, are an important part of your bird's diet, and either deficiencies or imbalances can cause health problems. Since both calcium and phosphorus are components of bone, a lack of either mineral can lead to weakened and diseased bones; so can an excess of phosphorus relative to calcium. A varied, whole-prey diet should provide your bird with the proper ratio.

In growing raptors, a lack of calcium or an unbalanced ratio of calcium to phosphorus can cause metabolic bone disease (rickets). Developing bones are weak, fragile, misshapen, and easily broken. Also, these birds often develop a curved spinal cord, causing an abnormal posture, balance and weight-bearing issues, and mobility problems. These are permanent abnormalities that cannot be reversed. Adult birds can also experience an improper ratio of calcium to phosphorus resulting in metabolic bone disease that predisposes them to pathological fractures. These changes, however, may be reversed with a good diet.

Water

Water is one of the basic necessities in life. In the wild, most raptors get a sufficient amount of moisture in the food they consume. If they don't, they will seek out an additional water source. In captivity, wild raptors are under stress, maintained on a diet that differs in water content from a wild, natural diet (freezing food decreases the moisture content) and are subjected to different environmental conditions, which may all affect the amount of water they need. Therefore, it is important to provide them with a bowl/bath of water so they have the resources to meet their water needs. However, if your bird has an injury that affects its balance, be careful to keep the water level shallow (chapter 4, Housing).

Raptors can become dehydrated if insufficient water is provided, they get overheated and lose excess moisture through heat dissipation mechanisms, their appetite decreases due to management changes or other stressors, or they develop an injury or illness. Dehydration can occur quickly and often leads to a cascade of medical problems.

7.3b Nutritional Excess

When we think of nutritional disorders, dietary deficiencies usually come to mind. Diseases resulting from too much of a particular nutrient also occur, however, and should be considered when establishing your bird's meal plan.

Protein

An "overdose" of protein can cause health problems in your bird, most commonly gout. Digesting high-protein diets, such as a diet of only organ meat (liver, heart, lungs), overworks the bird's kidneys; they are unable to get rid of all the nitrogen waste produced during protein metabolism. The nitrogen is converted to small crystals of uric acid that are deposited on the kidneys and other organs, resulting in organ failure. A whole prey diet consisting of a variety of prey items is important and will provide your bird with a healthy amount of protein. Each type of prey item has different protein content.

Fat

Low fat diets are often recommended for people wanting to lose a few extra pounds. As many people know from experience, eating an excess of fatty foods, although often tasty, can lead to obesity. Too much fat can cause atherosclerosis, a condition in which fat is deposited on blood vessels and organs, restricting or preventing their ability to function normally.

In raptors, a common potentially fatal disease resulting from obesity is Fatty Liver Disease (hepatic lipidosis). Fat infiltrates the liver, replacing liver cells and reducing liver function. To compound the problem, a bird with this disease often stops eating and an excess of fat is metabolized for energy, overwhelming the liver, which shuts down. Death is soon to follow.

If your bird is at a high weight and stops eating, take it to your veterinarian for diagnosis and treatment. Birds with fatty liver disease must be kept hydrated while put on a low fat diet (in severe cases, tubing a raptor with a low fat liquid carnivore diet may be necessary to help it recover), and can be provided with a metabolic aid such as lactulose to help the gastrointestinal system metabolize nutrients. To safely reduce the weight of an obese raptor (if tubing isn't necessary), put it on a reasonable diet (feed it a small amount of food every day), do not let it eat prey items that have an excessive amount of internal fat, and do not fast it to get the weight off — this may kill it by precipitating liver failure.

Vitamins and minerals

A few other nutritional excess disorders result from high intakes of vitamins or minerals. Believe it or not, too many vitamins are not good for either you or for your bird. Although illnesses from excessive vitamins are fairly rare in wild birds, they should be considered for captive birds, since you might plan on adding vitamin supplements to the diet. Make sure you follow instructions on the labels of any vitamins you use.

Vitamin D helps a bird absorb calcium from its diet and increases the amount of calcium in the blood, making the mineral available for functions that need calcium. However, an excess of vitamin D, most often resulting from an overdose of a vitamin supplement, can lead to deposition of blood calcium on the kidneys, heart, and other organs, decreasing their function. In addition, any unused vitamin D is stored in fat, and excessive amounts can be toxic. Supplemental vitamin D (cod liver oil) is often given to raptors. To avoid excess, however, it should be given in a small quantity only once per week to birds that are housed by themselves (see chapter 3, Diet). If multiple birds are housed together, do not provide cod liver oil — you cannot control how much each bird eats and an overdose can easily happen.

Phosphorus is a mineral that can also cause health problems if consumed out of proportion with calcium. Too much phosphorus (from a diet of meat without bones) can lead to bone disease. Not only does it stimulate the production of parathyroid hormone that draws calcium out of bones, it also binds to calcium in the intestines and prevents it from being absorbed into the blood. To prevent this, your bird's diet should have twice as much calcium as phosphorus (a 2:1 ratio) so that some calcium will be free to absorb into the blood. A whole-prey diet (meat and bones) should provide this ratio.

7.4 PARASITE INFECTIONS

If you're in need of a new conversation topic, here's one for you. Did you realize that throughout the world, more organisms are classified as parasites than non-parasites? Well, it's true, and as you might already know, wild and domestic animals (including humans) harbor a variety of them. Sometimes these organisms are relatively benign, creating only minor discomfort for their hosts. In other cases, however, parasitic infections lead to life-threatening health problems.

A parasite is an organism that lives on or in another organism, called a host, and usually causes it some degree of harm. It might rob its host of vital nutrition, produce and release toxins into the

Fig. 7.10
Common ectoparasites seen on wild and captive raptors (not to scale). (Gail Buhl)

(a) Feather Louse

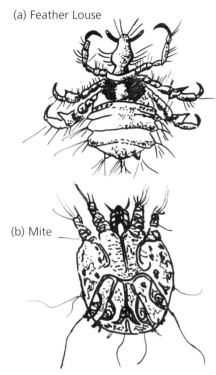

(b) Mite

(c) Tick

(d) Hippoboscid Fly

host's body, eat the host's body tissues or skin outgrowths (fur, feathers), or cause mechanical injury (boring holes into the host's body). Not pleasant creatures for cohabitation.

Parasites are also a diverse group of organisms. Some have a very specific host (and even a specific site on or within the host), while others invade a variety of hosts and can transfer from one to another. Some parasites even harbor other parasites. Parasites can be separated into three major groups:

- Ectoparasites, which live on the surface of an animal's body,
- Endoparasites, which invade the gastrointestinal tract, and
- Hemoparasites, which invade blood cells.

Parasites, in addition to being fascinating conversation pieces, are an important concern for every raptor manager. In the wild, most raptors carry a parasite load and are able to keep it in check unless they get sick or injured. You can expect that your bird will harbor parasites at some time during its stay with you. To keep your bird healthy, you need to learn about the most common raptor parasites and know what to do if you find them or suspect their presence.

7.4a Ectoparasites

Ectoparasites are usually easy to see and fairly simple to identify and treat. Four different kinds are frequently found on raptors (figure 7.10):

1. Feather lice are small, wingless insects, usually one to several millimeters long. They are generally classified as "chewers," feeding on feather vanes and the blood in quills of developing feathers. They are usually host-specific (that is, they won't live in your hair), some are species specific, and they are annoying enough that they can cause excessive preening and possible feather and skin damage.

2. Mites are tiny, spider-like parasites that live everywhere: on land, in fresh water, and in the oceans. In birds, they often infect feather follicles, the insides of feather quills, the skin, or the respiratory system. They are classified as "cutters" and can either cause disease themselves or create pathways through which other organisms can enter. Feather damage caused by mites is often seen around the eyes, legs, and feet of infected birds. Some birds can get a mite infection on their legs (scaly leg mites) causing feather loss (in birds with feathered tarsi), and crusty irritated skin.

3. Ticks are parasites that adhere to the skin. Like mites, ticks are classified as "cutters" and will create an entrance for other parasites and disease-causing organisms. Ticks can cause numerous health problems and often "bug" a variety of hosts. In raptors, ticks are most commonly seen on the eyelids or in the ears. Be aware that raptors can carry the ticks that cause Lyme disease, although few reports have been documented and the frequency of occurrence is unknown.

4. Hippoboscid flies are flat, bloodsucking flies that infect a variety of mammals and birds. In raptors, they embed themselves under the feathers and suck blood from the skin, often transmitting other parasites and infectious diseases into the blood.

Symptoms

Raptors often carry a light load of ectoparasites, especially if housed outdoors. If your bird is carrying a heavier load, however, you might find that it preens more frequently than normal. The other symptoms relate to the damage caused by the parasites. Feathers that appear to be chewed (figure 6.11c), owl feet that are devoid of feathers, or skin surfaces that look dry and crusty are good indications that your bird has an ectoparasite infection.

Diagnosis

Diagnosing ectoparasites is fairly easy. Most of these organisms can be seen crawling on the feathers (lice), attached to the skin (ticks), or jumping on and off your bird (hippoboscid flies). Mites can often be diagnosed by scraping infected skin or plucking a feather and viewing it under a microscope. If you find an ectoparasite and aren't sure what kind it is, you can either contact your veterinarian or place the ectoparasite on a glass slide in a drop of alcohol and view it under a microscope.

Treatment

Now that these fun-loving creatures have been identified, how can they be controlled? A variety of powders and sprays are available at pet stores or through your veterinarian. Many of the products have pyrethrins as the major ingredient. Always carefully read the label of any insecticide you use. Keep in mind that many parasite sprays are poisons, so follow directions carefully. It is possible for your bird to get sick if it either ingests a large amount of the insecticide when preening its feathers or inhales large amounts of dust/spray. So, apply as directed. Do not douse your bird, even if a heavy infection is present. You may need to apply the insecticide several times at three-to-four day intervals to eliminate the parasite load.

If you are using a spray, make sure you apply it from close range (within a few inches of the feathers) to avoid scattering. Keep it away from your bird's eyes. Most insecticide sprays are very irritating to eyes. If you need to apply it to your bird's head, spray some on a cotton ball or piece of gauze and wipe or massage in into the feathers. Also, when handling parasite sprays or powders, make sure to protect yourself. Pyrethrins are just as irritating to people as they are to birds. Make sure to protect your eyes, and wash your hands thoroughly after the application.

If mites are a recurring problem, they may be thriving in your bird's housing and re-infecting the bird. To eliminate them in the environment, you must thoroughly disinfect the housing. Soaking the ground, perches, water pans, etc., in a disinfectant such as a dilute bleach solution (5 to 10 percent) and then rinsing it thoroughly with water should do the trick (chapter 4, Housing). If not, an insecticide might be needed. Check with your veterinarian to find out what insecticide would be safe to use. If your parasite spray or powder is ineffective in eliminating mites on your bird, ivermectin given subcutaneously at a dose of 0.5mg/kg can also be effective.

Prevention

To keep your raptor free of external parasites, keep its housing clean (disinfect it regularly), and perform maintenance exams on a regular schedule (chapter 6, Maintenance Care). If you house more than one bird together, make sure that if one bird has a parasite problem, the other birds in the enclosure are also examined and treated.

7.4b Endoparasites

Endoparasites are a little trickier to identify and usually require microscopic examination of the bird's feces. Occasionally, adult worms will be passed, but usually the presence of endoparasites is diagnosed by eggs in the mutes (figure 7.11). There are several types of endoparasites frequently seen in raptors: roundworms (ascarids, *Capillaria*, and *serratospiculum*), protozoans (coccidia, *Trichomonas*), flukes, and tapeworms.

1. Roundworms are relatively long, narrow, smooth worms that are tapered at both ends. Three types are commonly seen in raptors: ascarids and *Capillaria*, both found in the stomachs and intestines of many species; and *Serratospiculum*, air-sac worms that are most common in prairie falcons. A fourth roundworm, *Syngamus trachea*

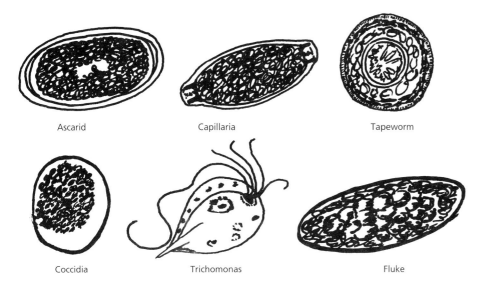

Fig. 7.11
Eggs of endoparasites commonly
seen in wild and captive raptors
(not to scale). (Gail Buhl)

Ascarid   Capillaria   Tapeworm

Coccidia   Trichomonas   Fluke

(gapeworm) has also been reported in some hawk species and common barn owls. Your bird could become infected with one or more of these types if it ingests contaminated food or water.

2. Tapeworms are long, segmented worms that inhabit the stomach and intestines. They have a small "head" at one end and taper at the other. Tapeworm segments, instead of eggs, are more commonly found in a fecal sample.

3. Protozoa are widespread parasites; two forms are most often found in raptors. Coccidia are small protozoa that can live in a bird's liver, digestive system, kidneys, and other tissues. *Trichomonas* is a protozoan of the digestive tract that often causes plaques and obstructions in the mouth and crop (figure 7.12). It is often seen in American kestrels, common barn owls, barred owls, Cooper's hawks, eastern screech owls, golden eagles, and peregrine falcons. Your bird could become infected with protozoa if it ingests contaminated food (feral pigeons, doves, starlings, and house sparrows are known to be carriers of *Trichomonas*).

4. Flukes are also known as trematodes. These parasitic worms vary greatly in size and shape and most commonly inhabit the intestines. Flukes seldom cause illness in a bird, unless a bird is otherwise debilitated.

Symptoms
Because you can't see endoparasites living inside your bird, you need

Fig. 7.12
Trichomonas in the mouth of an
adult sharp-shinned hawk.
(Photo by Dan Miller)

to watch for any clues that might reveal their presence. Occasionally, adult worms, especially tapeworms, are passed in the mutes — an easy clue. Usually, however, there are other symptoms that should make you suspicious. If your otherwise healthy bird starts having green mutes, pasty light brown mutes, or small amounts of blood in its mutes, that could indicate a parasitic infection. Either a change in your bird's appetite or an increase in the amount of food eaten with a subsequent loss of weight (without sudden changes in temperature, etc.) indicates a problem. In addition, keep in mind that changes in behavior, such as becoming listless and unwilling to cooperate, might be a clue that a bird isn't feeling well. Sick individuals are stressed and often develop heavy parasite infections.

Diagnosis

Many of these unwanted internal visitors can be diagnosed by examining a prepared fecal sample under a microscope. Often the parasites' eggs will be shed in the fecal matter; however, shedding might be periodic, so not every sample will contain eggs. If you suspect a parasitic infection, you might need to check several samples to find the culprit. Your veterinarian can perform the necessary examinations. If you are going to deliver a sample to your vet, make sure that it is as fresh as possible, and that you collect the dark center of the mutes, not the white urates.

Trichomonas — a protozoan parasite that creates plaques in the mouth or digestive system — can be positively identified by scraping the affected area (plaque), placing it on a slide with a drop of saline, and viewing it under a microscope. The organisms themselves are often visible. Your veterinarian can also do this test for you.

Treatment

Treatment for endoparasites depends on the type of parasite and must be prescribed by your veterinarian. Table 7.1 lists common drugs and dosages used to treat raptor parasites.

Prevention

To keep your bird "clean" of internal parasites, feed it only food that is from a known source and has been stored properly, keep its housing clean and disinfected, quarantine and examine any new birds to make sure they don't have transferable parasites, and conduct semiannual fecal examinations on your birds to identify and treat parasites that have sneaked in.

7.4c Hemoparasites

Hemoparasites, or blood parasites, can also occur in your raptor.

They are transmitted by biting insects and invade either red or white blood cells.

1. *Leukocytozoon* and *Haemoproteus*: The two most common blood parasites seen in raptors are of these genera. They can be transmitted by insects such as hippoboscid flies, louse flies, or biting midges. In an otherwise healthy bird, the number of these parasites is kept in check by the bird's immune system and they generally appear benign. Exceptions include nestling raptors in which *Leukocytozoon* are pathogenic (young birds do not have a fully developed immune system) and snowy owls in which *Hemoproteus* are pathogenic (snowy owls in the wild generally aren't exposed to this parasite and so their immune system is not prepared to deal with it). However, if any raptor becomes sick, these parasites can multiply and contribute to the bird's overall weakened state.

2. *Plasmodium*: This genus of blood parasite causes avian malaria, a disease that can be fatal if undetected and untreated. Avian malaria can be transmitted to your bird by a mosquito bite. Only certain species of mosquito (*Anopheles*) are frequent carriers of this parasite. If an infected mosquito bites your bird, *Plasmodium* organisms will enter the bird's bloodstream, travel to tissues such as the liver, and change into a different form that invades red blood cells and ruptures them. Therefore, one form of the organism invades tissues and another form invades the blood cells. To effectively treat a bird that has this disease, both forms must be killed.

Symptoms
In TRC's experience, almost any species can carry a light load of *Hemoproteus* or *Leukocytozoon* and not develop any clinical signs. Avian malaria is most commonly seen in northern raptor species, such as gyrfalcons and snowy owls, but has also been encountered in other species such as american kestrels, eastern screech owls, merlins, and northern saw-whet owls. If during the mosquito season your bird suddenly becomes lethargic and squinty-eyed, has bright (almost florescent) green mutes, loses its appetite, or drops weight, you should consider the possibility of avian malaria.

Diagnosis
Diagnosing blood parasites requires taking a blood sample and looking for the parasitic organisms in the blood.

Treatment

There is no treatment to eliminate *Hemoproteus* or *Leukocytozoon* from the blood. However, if a heavy load is present due to an illness, Trimethoprim/Sulfadiazine given orally at a dose of 30mg/kg twice a day for seven days can be given to treat the parasitemia and help parasite numbers return to a low tolerable level.

Avian malaria can be successfully treated if caught early enough. The current drug of choice is Mefloquin (Lariam) given at a dose of 30mg/kg at time zero, twelve hours, twenty-four hours, and forty-eight hours. It is important to medicate your bird at these times, as the drug needs to be in a high enough concentration to kill the blood and tissue forms of the parasite.

Prevention

We all know that in the summertime, mosquitoes and biting flies are prevalent. The safest way to avoid insect borne diseases is to house your bird in a mew covered with mosquito netting. This will help reduce the number of biting insects that have access to your bird and decrease the risk of these diseases. You can also bring your bird inside at night when biting insects are most active. If these options are not feasible, your bird can be given preventative drug therapy to prevent avian malaria. Mefloquin (Lariam) administered at a dose of 30mg/kg once per week during the mosquito season should help.

7.5 SUMMARY

There is no doubt that wild animals in captivity can pose a great challenge to their caretakers. As the primary manager of a raptor in captivity, it is your responsibility to make sure your bird remains healthy and injury-free. Regular maintenance examinations and observations of your bird will help detect problems so they can be addressed early.

Sometimes, however, birds become ill unexpectedly, exhibiting no symptoms until they are very sick.

Diseases such as aspergillosis, avian pox, and West Nile virus, parasite infections, nutritional disorders from an excess or deficiency in key nutrients, and management-related medical issues such as bumblefoot, lost talon sheaths, and frostbite are all potential ailments that can be encountered by raptors in captivity. If detected early, they can be successfully treated. If not detected early, they can be life threatening. Even if your bird only acts a little "off" one day, pay attention. It might be telling you that it has a problem. TRC cannot stress enough the importance of checking your raptor regularly to prevent life-threatening medical problems and correct management problems that might cause injury.

7.6 SUGGESTED READINGS

Beynon, P, N. Forbes, and N. Harcourt-Brown (eds.). 1996. *Manual of Raptors, Pigeons, and Waterfowl.* Ames, IA: Iowa State University Press.

Degernes, Laurel. 1990. Raptor Foot Care. *Journal of the Association of Avian Veterinarians.* Vol. 4, No. 2.

Forbes, N. and C. Flint. 2000. *Raptor Nutrition.* Evesham, UK: Honeybrook Farm Animal Foods.

Fowler, M. E. 1986. *Zoo and Wild Animal Medicine.* 2nd ed. Philadelphia: W.B. Saunders.

Gale, N. B. 1971. *Tuberculosis. Infectious and Parasitic Diseases of Wild Birds.* Ames, IA: Iowa State University Press.

Karlson, A. G. 1978. Avian Tuberculosis. *Mycobacterial Infections of Zoo Animals.* Washington DC: Smithsonian Institution Press.

Lumeij, J.T., D. Remple, P.T. Redig, M. Lierz, and J. Cooper. 2000. *Raptor Biomedicine III.* Zoological Education Network.

McCurnin, D. M. 1986. *Clinical Textbook for Veterinary Technicians.* Philadelphia: W.B. Saunders.

Redig, P.T. and J. Ackerman. 2000. Raptors. T.N. Tully, M.P.C. Lawton, and G. M. Dorrestein. *Avian Biology.* Educational and Professional Publishing, Ltd.

Redig, P.T., J. E. Cooper, J. D. Remple, and B. Hunter (eds.). 1993. *Raptor Biomedicine.* Minneapolis: University of Minnesota Press.

Remple. D.J. 2004. Intracellular Hematozoa of Raptors: A Review and Update. *Journal of Avian Medicine and Surgery* 18(2): 75-88.

Ritchie, B., G. Harrison, and L. Harrison. 1994. *Avian Medicine: Principles and Application.* Lake Worth, FL: Wingers Publishing.

Samour, J. 2000. *Avian Medicine.* London: Harcourt Publishers Limited.

Schmidt, G. and L. Roberts. 1985. *Foundations of Parasitology.* 2nd ed. St. Louis, MO: Times Mirror/Mosby College Publishing.

Steele, J. H., and M. M. Galton. 1971. Salmonellosis. *Infectious and Parasitic Diseases of Wild Birds.* Ames, IA: Iowa State University Press.

U.S. Dept. of the Interior Fish and Wildlife Service. 1987. *Field Guide to Wildlife Diseases.* Washington, DC: Resource Publication 167

Table 7.1 Drugs and their dosages for treating common endoparasitic infections in birds of prey

Parasite	Drug	Frequency	Dosage
Ascarids	Ivomec Panacur (10%)	Once Once, repeat in 10 days	0.2mg/kg 20-25mg/kg
Capallaria and Serratospiculum	Panacur (10%)	Once per day for five consecutive days; repeat with single dose 10 days after 5th dose. Do not give to young birds or turkey vultures	20-25mg/kg
Tapeworms	Droncit	Once, repeat in 14 days	30mg/kg
Coccidia	Albon Toltrazuril (baycox)	Once per day for 10 days Once per day for two days, repeat at day 14 and 15.	55mg/kg first day, then 25mg/kg 10mg/kg
Trichomonas	Metronidazole (flagyl) Carnidazole (spartrix – not currently available in the U.S.)	Once per day for 3 days (severe infections may require longer treatment) Once per day for 2 consecutive days (can be given up to 5 consecutive days for severe cases)	100mg/kg 20mg/kg
Flukes	Droncit	Once, repeat in 14 days	30mg/kg

Chapter 8: TRAINING

Training animals can be a very challenging, thought proviking process. It is not black and white, and the methods used to achieve a training goal depend in part on the goal itself. Thus, training a bird for use in education programs, free-flight demonstrations, or the sport of falconry often involves different approaches. Although many of the general concepts presented here apply to training raptors in general, the specific methods listed are for training permanently disabled raptors for program and display use.

There are so many things involved in handling and training an educational raptor, along with so many opinions on the subject, that it could easily be a book in itself. In this chapter, basic training guidelines will be presented to help you get started.As mentioned previously, there are several approaches used to train a raptor. Which one you choose will depend on your experience level, the way in which your bird will be used, the natural history of the species you are training, and the bird's individual personality. Most often, the bird will guide the way.

The topics that will be discussed include training philosophies and goals, preparing for your bird's arrival, what to do when your bird first arrives, communication between you and your bird, basic training techniques, and introducing your bird to groups of people.

8.1 TRAINING GOALS

You just received a crate with a rehabilitated wild raptor in it. This raptor is going to be your responsibility for the next five to fifty years. What do you do now? Perhaps you are interested in a display-only bird, but maybe you want to jump in with both feet and train a raptor for education programs. The first question to ask is, "why?" If the answer is because it is cool, that is not enough of a reason.

For many groups, the role of an educational raptor is a stepping-stone to get the general public interested in wildlife and wildlife conservation. For thousands of years raptors and birds in general have fascinated humans. These amazing creatures can

Fig. 8.1

A well-trained bird perched comfortably on the handler's fist during an education program.

inspire children and adults alike to learn more about wildlife and practice responsible environmental stewardship. You can only inspire when your bird is properly trained and comfortable in the situations in which you present it. Bringing a stressed, unhealthy bird in front of the public is not only uninspiring, but also irresponsible and inhumane.

An important component to any training program is to set achievable goals. If you set your goals at an unreasonable level that your raptor cannot reach, you and the raptor will be frustrated. Think about what you want your bird to be able to do. Is it something the bird is capable of? If so, define this as your goal.

Always take into account the species' natural history. For example, it would be much easier to train a woodpecker to peck on a log for food, than to train it to dive into water to capture fish. This extreme example can be extrapolated to the raptor species you are working with. A screech owl is more likely to sit quietly on the glove than a peregrine falcon is. The reason is that when threatened, one of the screech owl's means of defense is camouflage. It doesn't want to draw attention to itself by moving. The natural reaction of a peregrine is to fly away.

Knowing the natural behaviors and responses of a species will help you set reasonable training goals and choose the training method that is most compatible. For instance, a non-native species, the hooded vulture (*Necrosyrtes monachus*), plays dead when stressed or threatened. This type of behavior is important to know before training a hooded vulture. Imagine your shock when the bird does not stand on your glove but goes rather limp and rolls its head back on its body!

The Raptor Center's basic training goals are to have a bird step calmly onto a handler's glove and stand without bating during an education program (figure 8.1). The bird must also perch steadily on the glove during movement to and from its mew, the program room, or the training area; to step calmly back into its crate or into its mew; and calmly step on and off a scale for weighing. A bating, screaming, stressed-out bird is not acceptable. By carefully training your bird and reacting to its body language you can set your bird up to succeed. It is not unreasonable to require a raptor to sit calmly on a glove before it can be added to a program. In fact, it is a must.

8.2 PREPARATION

Once you have decided to make the commitment of caring for a captive raptor, there are several things you should do to familiarize yourself with appropriate training and management protocols before you acquire a bird. It is important to take the time necessary to thoroughly prepare yourself. Don't be surprised if this takes six months to a year. Raptors can be high maintenance animals and there is a lot to know about training them and keeping them healthy in captivity.

8.2a Training

Whether you realize it or not, once the bird reaches your doorstep, its training begins (and yours does too), therefore, it is important to have a training plan laid out before your bird arrives. There are several resources available to help you decide on a plan and provide you with the framework you will need for success.

- First, it is a good idea to visit facilities (zoos, nature centers, rehabilitation facilities) that manage birds for display and/or program use. Talk to the handler(s) and caretaker(s); take notes about housing, management, and handling techniques, assess the birds' overall health (beak condition, feather condition) and general attitude (calm, nervous), and determine whether the birds behave in a manner consistent with your goals.

- Second, it is strongly recommended to shadow an educator or falconer to observe handling and management techniques, learn how to interpret bird behavior, and advance your knowledge of techniques available to train your avian educator in its new role. An ideal situation is to get a knowledgeable person to be your mentor and assist you in management and training when you do acquire your bird.

- Third, some experienced individuals/facilities offer workshops on training/managing captive raptors. These include The Raptor Center, College of Veterinary Medicine at the University of Minnesota, the World Bird Sanctuary, Natural Encounters, Inc., and the International Association of Avian Trainers and Educators. It is a relatively small investment to make for the value of having a well-trained, healthy educational bird contributing to your program for many years.

- Finally, good books and videos on general animal training principles are available. They can help you shape your behaviors so you are communicating clearly and consistently with your bird. This is a very critical component of successful training (see suggested readings at the end of this chapter).

8.2b Management

Once you identify a specific bird, make sure to find out its previous management history. Important things to know include the bird's weight range, diet (type and quantity of food), housing situation (convalescent cage, indoor free-loft, outdoor free-loft, housed with other birds or solo), type of perches it has been offered, and the climate it has been most recently exposed to. These things will give you the groundwork for preparing for your new educator's arrival.

When your bird does arrive, its permanent housing should be completed and ready for occupancy and you should have an adequate supply of appropriate food items. If you are going to use your bird on the glove for programs, equipment should be made and be ready to fit and apply. The clear communication path begins immediately when you acquire a bird and you should start by letting it know what the "rules" are in its new environment. For example, it is not recommended to free loft a new bird for a few days before applying equipment and tethering it in the mew. Let your bird know right away that when it is in the mew, it will be tethered. If you are inconsistent in your messages, your bird will be inconsistent in its behavior.

8.3 YOUR BIRD ARRIVES: WHAT DO YOU DO?

Okay. You have done all your homework and are ready to acquire your bird. What is next? First, make sure that you arrange for a convenient time to receive your bird. Don't agree to get a bird when you are rushed, preoccupied, do not have someone to assist you with an exam and applying equipment, or have little time to get it settled. Next, if you are going to pick up your bird and transport it to its new home, you will need an appropriate-sized carrier (chapter 9, Transporting Raptors). Now you are ready to begin your new adventure.

8.3a Conduct an Exam

It is highly recommended to examine your new bird so you know its initial general condition and can identify possible concerns. You could take your bird to your veterinarian immediately upon arrival to conduct a complete exam or do a brief exam yourself. However, if you examine it yourself, TRC strongly recommends making an appointment with your vet within one week of your bird's arrival for a more complete exam (including basic blood work and a fecal exam). To conduct a brief exam, you will need an assistant, basic handling equipment (gloves, protective eye wear, a flat padded surface, a towel or hood to cover the bird's eyes), and paper or a basic exam form to record your findings.

1. Observe the bird in the carrier. If it has a physical disability, pay particular attention to the affected site. If a wing droops, how far down does it hang? How is it held relative to the body? How do the tips of the primary flight feathers rest over the back? If the bird had a leg injury, note how the bird stands — Does it lean? Does it bear weight evenly on both legs? If the bird suffered eye damage, look at the condition of the affected eye — Is it sunken, discolored, having an abnormal pupil shape, or missing altogether? These types of observations are important so you have a basis of where your bird is starting from and can assess if changes occur. You can observe the affected area more closely when you conduct a basic exam.

2. Remove your bird from the carrier. This is often a very traumatic experience for both the bird and the new handler. Unless you are acquiring a previously trained bird, a bird will not willingly come out of the crate for you. So,

you will need to carefully grab it. It is recommended to have an assistant be the "bad guy" and do the grabbing. TRC recommends doing a body grab if possible (chapter 6, Maintenance Care). However, be careful. Keep in mind that the bird you are getting is not a tame animal and it will readily defend itself with its beak and talons. You may receive a bird that already has anklets and jesses applied. Do not attempt to carry the bird out on your gloved hand unless it has been previously trained. A naïve bird will not have a clue as to what you are trying to do and it will struggle greatly in the crate, potentially injuring itself. Remember, in training, you only want to "ask" your bird to do something when there is a reasonable chance it will succeed.

3. Record the bird's weight. Once the bird is out of the crate, restrain it in a cradle position and weigh it (chapter 6, Maintenance care). One of the most important pieces of information to receive on a new bird is its weight. This will be a critical reference number in the future.

4. Restrain the bird on a table and conduct a brief physical exam. Place the bird on its back on a padded table, restrain it safely (chapter 6, Maintenance Care), and cover its head to reduce its stress. Then, examine and record the condition of the following:

    A. Flight, tail, and body feathers. Are they broken, tipped, in blood, missing, or chewed by parasites?

    B. Beak. Is it at a proper length, thickness, and shape or does it need coping?

    C. Feet. Do the bottoms of the feet and toes look healthy or do they have pink, thin, scabby, or abraded areas?

    D. Wrists. Are the wrists abraded due to the bird bouncing around a confined space?

Broken blood feathers, an overgrown beak, unhealthy feet, and damaged wrists should all be taken care of as soon as possible. Do not attempt to begin training or acclimating a bird to a display until all medical issues are resolved (chapter 7, Medical Care). An overgrown beak can usually be fixed quickly, but the others may take some time. Ideally, these issues would have been identified and resolved before you acquire a bird (chapter 2, Selecting a Bird for Education). You want the training process to be positive from day one. The care involved in treating medical problems is not a positive experience for a bird.

Also, if your bird has a permanent physical injury, examine the disabled area to know exactly what the situation is. Take a photograph of it so you have a record of your starting point and can better assess if changes occur. Questions to answer may include: Are feathers missing over an old fracture site? Does a wing or leg have an abnormal bend to it? Does one eye look different than the other? Identify your bird's individual physical characteristics so you can monitor them for changes.

8.3b Apply Equipment

If your bird will be used for program use, once you have finished the exam and feel that the bird is healthy and ready to begin manning, apply the equipment (chapter 5, Equipment). It is a good time to do this when you already have the bird restrained on the table. Get as many negative encounters as possible out of the way that first day.

8.3c Transfer to Housing

Unless you are going to man your bird with traditional techniques (see below) and are ready to begin immediately, the next step is to place your bird in a housing environment. For display birds, this means introducing your bird to its free-loft mew. For program birds, it is best to tether them in a quiet training area, away from public viewing. Do not attempt to get your bird on the glove to transfer it to its housing, unless you are beginning the traditional manning process as described later.

Display birds

There are several approaches you can take to acclimate your bird to a new display. Before you do so, make sure the mew is ready. If the display has any glass walls or windows, these should be marked with an interrupting pattern to prevent the bird from crashing into them. Streamers, colored tape straps, or other patterns securely placed on the outside (or inside if there is a lot of glare) will help break up the image of the glass and alert the bird that it is a solid structure. Once a bird adapts to the glass, the distracting devices can be removed a few at a time until you are sure your bird is aware of the hard glass surface.

Time of day: The best time of day to introduce your bird to its new home will depend if it is a diurnal or nocturnal species, and if it is going into an empty display or one that already houses additional birds. If your bird is nocturnal, it is a good idea to put it in the display during the day, when it typically is less active. Keep an eye on it and monitor its behavior. If it is very active and nervous, it may be a good idea to put it in a carrier at night and the display during the day until it gets more accustomed to its new surroundings. Nocturnal birds are by nature active at night and if a bird is already nervous and active during the day, who knows what will happen during the moonlit hours?

If the bird is a diurnal species and is going into an empty display, you can either place it in the display during the day and monitor it frequently to see how it is adapting, or put it in at dusk when it will be calm and let it wake up to its new surroundings in the morning. If other birds are present in the display (see chapter 4, Housing), you should add the new tenant during the day when you can observe bird interactions and intervene if necessary.

All birds new to a display should be monitored closely until they appear calm, are eating well, and if other birds are present, handle the companionship without problems. Birds that are intimidated by the presence of other birds often won't eat and either move excessively or remain stoic unless provoked. Also, when you add a bird to an occupied display, pay attention to the interactions between the birds. If there is any aggression (chasing, vocalization, or full out attacks), remove the new bird immediately. Aggression usually occurs from the resident who is territorial of its space. You may need to house the birds separately or move both of them to a new mew, one where the resident bird has not established a territory. For more information on housing birds together, see chapter 4, Housing.

How to place your bird in a display: There are two major methods of introducing your bird to a display setting. The first can be called a "hard" introduction, where you place the bird directly in the display during the day. The bird is immediately confronted with a new space and environment and is forced to adjust from that point on. The second method can be called a "soft introduction." This involves placing your bird in a carrier, placing the carrier in the display, slowly opening the carrier door, and then leaving. The bird can leave the carrier when it feels comfortable (sometimes it will bolt out right away) and will generally be a little less panicky. You can also leave the bird in a crate for a few days in the display before opening the door. This approach is not recommended, however, if other birds are present in the display.

Program Birds

There are a few different approaches to the initial training of a program bird and which one you choose will determine how to house your bird from day one.

As you will read later, the traditional training approach "forces" the bird to perch on your glove immediately and then a great amount of time is spent with the bird to get it accustomed to the glove. If training by this method, it is recommended to house a diurnal bird in the dark for a few days between manning sessions, either by tethering it with a hood (see chapter 4, Housing) or keeping it in a carrier that has a wide door for smooth entrances and exits. The most stressful part of this type of training is getting the bird on your glove, and for a diurnal bird it will be easier to do so if the bird can't see you.

The modern training approach uses operant conditioning to teach your bird what behaviors you want; you do not force the bird to do anything. It takes a little longer to get the bird on your glove, but it is a more positive experience and less bating generally occurs. If training by this method, it is recommended to house the bird by tethering it in a training area to keep it calmer. Then, only the trainer will have direct access to the bird and the bird won't receive mixed messages. More details on operant conditioning will follow.

8.4 COMMUNICATION

Clear communication is an essential component of a successful training program. Your behavior will greatly influence your bird's behavior, and in turn, your bird's behavior will influence how you alter your behavior to get the desired results. The challenging part can be to interpret what your bird is trying to tell you. The key is body language.

8.4a Body Language: A Glimpse into a Bird's Mind

It's pretty obvious that a bird can't tell us verbally what it is thinking or feeling — but that doesn't mean it can't communicate with us. Like humans, birds communicate effectively through body language. It's your job to observe your bird and its body movements and learn what it is telling you. This is critical in the training process, as it will help you predict and control the bird's actions.

To help you learn this new language, we are going to provide you with a few examples of the most common messages birds give us through their body movements.

Fig. 8.2
A bald eagle perched on a glove exhibiting signs of stress.

Fig. 8.3
Common signs of aggression in captive raptors.
(Gail Buhl)

(a)
A bald eagle lifts its "hackles" to show annoyance and alarm.

Signs of stress

To be a responsible raptor caretaker, you need to be able to identify when your bird is uncomfortable or stressed so you can improve the situation and provide relief. Signs of stress include raised hackles, excessive bating, panting or gular fluttering, droopy wings, refusing to stand, and in some hawk and small falcon species, open mouths with their tongues protruding out (figure 8.2).

The raised hackles and open mouth should subside in time with exposure to the stressor, as long as nothing negative occurs. However, if your bird exhibits the more severe signs of stress (excessive panting, droopy wings, refusal to stand, constant bating) remove it from the situation immediately and monitor the bird closely as it recovers. These behaviors are an indicator that the bird is not just mentally stressed, but physically stressed as well, which could lead to serious health problems. A bird in this state was pushed too far too fast and was clearly not ready for the situation it was placed in.

Example: Your owl is on your gloved hand, and this time you are presenting it to a small group of children. After a few minutes, the bird starts open-mouth breathing, and the feathers under its chin move in and out (gular fluttering). What's going on?

a. Your bird is becoming stressed and hot.
b. Your bird is sick and about to vomit.
c. Your bird is expanding its throat patch to catch bugs.

The correct answer is (a). When an owl becomes stressed or hot, it begins to flutter its throat patch to loose heat. A hawk will pant (similar to a dog) by opening its mouth and increasing its breathing rate. All species of raptors droop both wings when overheated. If the room is cool and your bird is exhibiting these behaviors that means it's becoming extremely stressed with the situation and should be removed.

Signs of aggression

At some point in its stay with you, your bird will probably demonstrate aggressive behaviors. These behaviors may be directed towards you, another person, another bird, or something potentially threatening in its environment. Signs of aggression include raised hackles (figure 8.3a); deliberate footing (figure 8.3b); an increase in the bird's tension (gripping firmly on the fist, tense body); or biting (figure 8.3c). Just like us, raptors can have different moods, and their behavior can be influenced by weather, time of year, and stress level. Even the most even-tempered bird can have "grumpy" days.

Example: Imagine that your hawk is perched on your gloved hand. When you lift your opposite arm and hand, the feathers on the back of its neck (its hackles) are raised (figure8.3a) and the bird stares at your arm. Which of the following explains this behavior?

a. Your bird is erecting its feathers to cool its head.
b. Your bird is curious about this large moving thing that is attached to you.
c. Your bird is alarmed and becoming defensive.

If you answered (c), you're correct. A hawk will raise its hackles if confronted with a situation it perceives to be potentially dangerous or threatening. It's a defensive reaction. Hawks also communicate with each other in this way, raising their hackles to show ownership of a territory or a meal. This is usually accompanied by spread wings.

If your bird raises its hackles in response to something you do or a situation in which you put it, what should you do? You can either remove your bird from the stimulus, or expose it to the stimulus more often (every day) so that it becomes "desensitized" and accepts it without alarm. You might find that when you begin training and acclimating your hawk to its routine, its hackles will be raised often (even upon the sight of you).

If presented with the same situation, an owl might raise all its feathers and open its wings in a cupped fashion, to increase its apparent body size and look threatening to the intruding object or person. It usually also sways and stomps back and forth and hisses or clacks its beak. This is a "fight" response. Once your owl has been exposed to more situations, if it feels threatened or otherwise uncomfortable, it might rapidly move its head back and forth (not because it wants to exercise its neck muscles) to look for the closest exit (a "flight" response). This latter behavior usually is followed by a bate (an attempt to fly off the fist). Either removing the stimulus or turning your body or arm slightly to disrupt your bird's balance might help to settle the bird down.

Some of the stimuli that often create these defensive responses include dogs, cats, hats, people with specific qualities (believe it or not, some birds prefer women over men, or prefer men without beards), colors (especially for hawks), balloons, things moving on wheels (bicycles, wheel barrows), or people in general if the bird is broody (exhibiting nesting behavior) or a human imprint.

Signs of distraction

When trying to train animals, it is important to have their undivided attention. It is often challenging enough to convey clear messages without the trainer or trainee being distracted.

(b)
A red-tailed hawk aggressively foots the handler's glove.

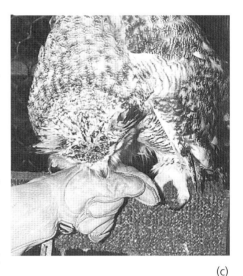

(c)
A great horned owl bites the handler's glove when trying to be fisted.

Fig. 8.4
A rouse in a merlin (Gail Buhl)

Example: Your great horned owl is perched on your gloved hand outside. All of a sudden, its starts moving its head up and down and back and forth repeatedly. What's going on?

    a. Your bird has something stuck in its throat.
    b. Your bird is trying to find something it hears in the distance.
    c. Your bird has a stiff neck and is trying to work it out.

Congratulations if you answered (b). An owl focuses on objects it hears by moving its head until it faces the direction of a sound (this is partly how it hunts in the wild). Owls have wide heads and several species have ears that are placed asymmetrically on the sides of their heads. Due to these features, sounds arrive at different times to each ear. This timing difference allows the owl to localize a sound by moving its head until it faces the sound directly. So, if your owl is moving its head in this fashion, something is catching its attention. Usually this makes an interesting behavior to talk about during a program. However, if you're training your owl to step onto a scale, enter a travel carrier, etc., and its attention is elsewhere, your training will go nowhere. Either remove the stimulus or let the owl find the sound and satisfy its curiosity before continuing training.

    Falcons, in contrast, bob their heads up and down when they see something they want to look at more clearly, and hawks tend to stretch their necks high and out. Be aware that these behaviors indicate that your bird's attention is not on you or your training, and that the bird might soon attempt to fly either toward the stimulus or away from it. Removing the stimulus from view may prevent these behaviors.

Signs of comfort

Up to now, we've talked only about behaviors that represent stress, discomfort, aggression, or distraction, all undesirable behaviors. Don't worry, birds can also tell us that they are comfortable.

    Example: You are presenting your bird to a group of school children and it is perched calmly on your gloved hand. After about five minutes, it lifts up one foot and tucks it into its body. This means:

    a. Your bird's foot is sore and it needs to rest it.
    b. Your bird is comfortable enough to stand in a normal roosting position.
    c. Your bird is trying to hide its foot to fool the children.

Hopefully, you answered (b). One of the signs of comfort in a bird is when it tucks up a foot, as it does when roosting. It feels like it is safe and does not need to be on guard. If something causes concern, it will immediately put the foot down, in anticipation of needing to move quickly.

A few additional signs of comfort are if a bird begins preening, rouses its feathers (lifts them all and then shakes them back into place) (figure 8.4), rests a leg on its hock (ankle) joint, or bathes. These actions take part of a bird's attention off its surroundings and it will only do them if it is feeling secure. Be patient. It will take time and handling to get your bird to this level.

The bottom line in understanding bird behavior is to take the time to evaluate what is stimulating a response from your bird. Then, if the behavior is undesirable, think about a way to avert it by distracting the bird or removing the stimulus. Also, keep in mind that your body language can, and often is, one of the contributing stimuli. A bird perched on your gloved hand can sense if you are tense, about to move, or hurried. Even slight movements on your part might put the bird in an uncomfortable position and cause a reaction. Never be in a rush when handling a raptor. Being relaxed and comfortable holding your bird, and offering it a stable, solid perch on your glove, will go a long way to teaching your bird that your glove is a safe place to be.

8.4b Human Behavior around a Raptor: A Glimpse into Cause and Effect

Before we talk about specific methods for training your raptor, there are a few basic guidelines about your behavior that we want to share. Keep in mind that the speed and effectiveness of the training depends to a large extent on your behavior and attitude. You'll learn and improve with experience.

1. Always move slowly but confidently around your bird. Quick, jerky movements are unpredictable and might cause your bird to try to fly away from you. All motions should be smooth and continuous.

2. Be consistent. Always use the same glove on the same hand. Develop a routine such as gloving–weighing–walking –returning to mew–feeding. Once a bird expects a routine, alterations can be made and accepted more readily. If you are not consistent, especially in early training, your bird will be more nervous and reactive and the training will take much longer.

3. Work with your bird to get it used to your hands. Slowly

move your free hand in front of it, touch its feet, and so on, so it gets used to the close movement. That way, it won't be alarmed as you move your hand for other purposes or attempt to lift up the bird's foot to check it. However, be careful and watch your bird. If it stares at what you're doing and lowers its head, raises it hackles, or attempts to bite, slowly retreat a little. A few seconds of hand movements at a time is all that's necessary.

4. Be patient with your bird, and don't try to rush its actions. Arrange your handling sessions when you aren't short of time or feeling pressed to get other things done. Through your body actions, raptors can sense your energy level. If you are rushed and anxious, the bird will be more active.

5. Monitor other people's behavior around your bird. If you have your bird on your gloved hand or tethered for display, make sure that people keep a respectful distance so the bird doesn't become nervous; even birds need their own personal space. Also, don't allow people to completely encircle the bird. Instead, have them stand or sit in front of it or to one side. If the bird sees people on all sides, it will be nervous and constantly scan all directions looking for danger. If you are walking with your bird and people are accompanying you, have them walk either in front of you or on the side opposite from the bird. That way, your bird will be able to see everyone and won't be as edgy. Respect your bird by providing it with a comfortable space to perch and be observed.

8.5 TRAINING METHODS

As you can see, a great amount of preparation is necessary before you and your bird are ready for training. You'll find that both you and your bird will be students, training each other. Communication is a two-way street and to create solid, dependable behaviors, clear communication, repetition, and consistency are essential.

As mentioned earlier, there are numerous opinions on the most effective way to train a raptor. What TRC presents here is a foundation to get you started. Other raptor handlers might train birds differently, with great success. Contact experienced individuals to get their opinion on how to get the job done. Inevitably, your bird will let you know how you are doing and whether your strategy needs to be altered. No one can tell you how to train your bird better than the bird itself. You just need to know how to listen.

8.5a Traditional Method

The traditional approach to training a raptor is the least complicated and basically forces a bird to accept situations right away. A new bird is placed on a gloved hand and held for long periods of time. It will be nervous and might bate frequently, but eventually it will get tired and resolve to being on the glove.

This type of training is an old falconry technique called "waking." Falconers would have their birds on their gloved hand for twenty-four to forty-eight hours straight and both parties would be quite mellow at the end due to exhaustion. However, for practicality, a new education bird is usually held on the glove for only one to two hours at a time. With this type of training, the bird has no control over the process.

Many owl species, which are calmer, do not want to draw attention to themselves and are most comfortable when they blend in with the environment; they can still be successfully manned in this fashion.

8.5b Modern Method: Operant Conditioning

The modern and more analytical approach to training involves a type of behavioral modification called operant conditioning. The main theory behind this method is that a bird has control of its own training and really leads each session. It is not forced to do anything and can choose to participate or not to participate. In order for this to work, a bird must be motivated to do a certain behavior and then rewarded with positive reinforcement when it completes the behavior. Motivation and reinforcement are the two factors that will make a bird choose to participate in the first place. Let's take a closer look at these two key components.

Motivation

Training is all about motivation. There is a reason for every action. Always think about any behavior, whether positive or negative, in terms of what the motivation was for a particular behavior. For example: what is the falcon's motivation to step onto your glove and stay there calmly for ten to fifteen minutes? There has to be something positive that the bird receives when on your glove, or some reason that your glove is better for the bird than where it was before. For a raptor there are two main motivators: food and safety. There are also other weaker motivators that you might not notice. These include things like sunlight for sunning, water for bathing, moving from a noisy area, etc.

Many behaviors that raptors exhibit have been refined over

time to be advantageous for the species. For example, aggressive displays make a predatory animal leave. If a behavior were not advantageous, it would eventually cease to exist. Your bird will often react to you and the situations in which you put it, in the same way it would react to life and death situations in the wild.

For training raptors, one of the positive motivators of choice is food. This works best with diurnal birds, as most nocturnal birds are more comfortable eating during the evening. Your bird's weight does not need to be lowered to a critical level in order for it to step to your glove for food. Nutrition is a basic need for survival and if a bird learns that you fill the need, it will anticipate your arrival. Once a bird makes that association and steps up nicely onto the glove, the behavior of stepping up can be reinforced intermittently (you will not need to feed it every time to keep the behavior strong).

Reinforcement

Reinforcing behaviors increases the likelihood they will happen again. This can either be done with positive reinforcement — the type of choice for training birds — or negative reinforcement. Positive reinforcement is something an animal wants, such as food, security, tactile stimulation, praise, etc. and is presented at the time when a desirable act occurs. It is an immediate reward for good behavior. As mentioned above, food and safety are two reinforcements that can be used to shape a raptor's behavior. The tactile stimulation (petting) that many mammals understand as a positive reinforcer does not work with raptors. It also contradicts the message that raptors are not pets.

Negative reinforcement, on the other hand, is something undesirable that an animal wants to avoid. For example, applying pressure to a dog's rear end will increase the likelihood that the dog will sit. The dog learns to sit (when given a cue — see 8.5b-3 below) to avoid the pressure, which is unpleasant.

The most effective way to train animals (and people too) is through positive reinforcement. Think of your own experiences. Do you enjoy learning in a situation where you fail frequently and the teacher reprimands you, or are you much more eager to learn when you are set up to succeed and are rewarded for your successes?

When training your new educator, your job is to set the bird up to succeed. This is not as easy as it sounds. You need to break each behavior down into baby steps. Don't ask a tethered bird to step on a gloved hand, be removed from its mew, walk through a doorway, and get on the scale without bating the first day. Just stepping on the glove calmly and willingly is a huge first step. This may take a while and should be repeated consistently before moving on to walking, going through doorways, etc. Each behavior should be consistent and solid before moving on to the next behavior. If the bird is pushed too hard, too fast, it may regress; instead of stepping up onto the glove, it may bate away from you in a panic. This is not positive reinforcement.

A trainer usually has one of two responses to an animal's behavior: the desire to eliminate it or the desire to reinforce it. There are many different ways to extinguish or reinforce a behavior. To get rid of undesirable behaviors, many old training techniques involve punishment, which is something that happens after the action occurred and is something the animal cannot avoid.

For example, you come home from work and your dog has gotten into the garbage. You call the dog, stick its nose in the garbage and yell at it. Will this prevent the dog from doing it again? More than likely, it will not. Animals live in the present and the dog was no doubt highly rewarded at the time by eating some juicy leftovers. The punishment came late, did not prevent the behavior, and in this case probably won't override the reinforcement the dog got by eating the garbage. You are much better off to hide and/or secure your garbage in the first place.

Punishment should be avoided, even if provided immediately following an action, as animals will only act at the minimum level required to avoid it.

In operant conditioning, a behavior is eliminated either by not providing positive reinforcement or providing negative reinforcement (such as the pressure applied to a dog's rear to get it to sit or spraying an unpleasant taste, like lemon juice, on the garbage). If a bird is not somehow reinforced to behave in a certain way, it will stop that behavior. As a trainer, the key is deciding what your bird considers reinforcing.

As already stated, motivation and positive reinforcement are two critical components of operant conditioning and understanding them will help get you started in training your bird. You might be thinking to yourself, "How do I ask for a specific behavior"? Or "How do I make sure not to reinforce the wrong behavior?" These questions can be answered by defining a few more operant conditioning tools.

Cue

In order for a bird to perform a certain behavior, it needs to know what the trainer is asking for. A trainer can communicate this to a bird by associating a sound or action with a desired behavior. In the case of stepping up to the glove, a glove presented slowly at foot level is a cue that the trainer wants the bird to step up. A crate with an open door can cue a bird that the trainer wants it to enter.

Cues are also used regularly in dog obedience training. When a puppy sits on its own, the owner praises it and says, "good sit!" This praise is usually accompanied by a yummy dog treat. The dog soon makes the association between the word "sit" (the cue) the act of sitting, and the praise or treat. When a cue is associated with positive reinforcement, it then becomes a conditioned stimulus.

Conditioned stimulus

A conditioned stimulus is another way that you can communicate with your bird. It is a sound or action that an animal has been repeatedly exposed to and is associated with positive reinforcement. For example, the presentation of a gloved hand cues the bird to step up and the act is followed by the presentation of a tidbit of food, movement to a quieter area, etc. A click or whistle is also sometimes used as a cue to ask a bird to step on or off the glove, into a crate, on and off a scale, etc. Once the behavior occurs, the animal is rewarded and learns to associate the click or whistle with the reward.

Bridge

Just as the name implies, a bridge (a type of conditioned reinforcer) is a sound or action that immediately tells a bird it performed the desirable behavior and fills the gap of time between the act and the positive reinforcement. Timing is critical in training. You need to let the bird know it was successful immediately after the behavior occurred. This can be tricky because animals react so quickly that even in the few seconds it takes to offer positive reinforcement, another behavior could be slipped in and be the actual recipient of the reward. A bridge makes sure the right behavior is reinforced and lets the bird know that the real reward is coming.

For example, your bird is perched calmly on your glove in a new area and you want to reward it by giving it a juicy piece of food. As you reach into your "treat" bag, the bird senses your movement and bates just as you bring the tidbit up to the glove. If you let the bird have the tidbit, you have just rewarded the bate and not the act of sitting quietly. You could have given the bird a click or whistle to "praise" the behavior of perching quietly and then follow it up with a tidbit.

Once you and your bird have identified a bridge through repetition, using it to reinforce good behaviors will allow your bird to progress rapidly through its training program. The two of you will have created a form of clear, effective communication. In the example above, if you made a click before reaching into your bag, the bird would have anticipated the reward and would probably not have bated in the first place.

Clearly, training a bird using operant conditioning requires a working knowledge of the theories behind it. It may seem a little complicated, but people use it successfully every day and probably aren't even aware of it. In fact, you have no doubt used it without even knowing it. If you are going to train your bird with this technique, work one behavior at a time. Define the behavior you want to train and work the training process out in your head. Here are some questions you should ask:

1. Are there baby steps involved?
2. What am I going to use to cue the behavior?
3. Am I going to use a bridging reinforcer and if so, what will it be?
4. What type of positive reinforcement will motivate the bird?
5. What undesirable behaviors should I make sure not to reinforce?

It may seem like a lot to identify for each behavior but once you get the hang of it, you will move quickly through the training process. These techniques work well on people too, so before you use them on your new bird, try some positive reinforcement on your family and friends and see how it works. You will learn that timing is everything.

8.6 TRAINING THE TRAINER

Before you start training your bird you need to know some basic handling techniques, such as how to properly hold your bird on the glove, and what to do when it attempts to fly off the fist (bate). Understanding and practicing good handling techniques from day one will make the training process smoother for everyone.

8.6a Holding a Bird on the Glove

Raptor handling 101 begins with a lesson on the proper posture for holding your bird. With your arm slightly out from your body (most people hold a bird on their left hand), bend your elbow and make a fist (your fist should be about level with your chest). Position your hand so that the index finger is on top; this part of your fist provides a solid surface for the bird to perch on (figure 8.5). Make sure that your fist doesn't drop lower than your forearm. If it does, the bird will try to walk up your arm (birds like to perch on the highest point). Once you feel comfortable

Fig. 8.5
Proper position for holding a raptor on the gloved hand, illustrating the jess lock and looped leash. (Ron Winch)

Fig. 8.6
Agitated red-tailed hawk attempts to get a little slack from jesses held too tightly. (Ron Winch)

Fig. 8.8
Using gentle hand movement to help a great horned owl recover from a bate. (Gail Buhl)

holding the bird in this fashion, you can offer it greater security by bringing your arm closer to your body. Do this only after your bird accepts you and you feel confident that it won't try to bite your face or body.

Now pretend the bird has stepped onto your gloved hand (this will be addressed below). Place the jesses between your thumb and index fingers and then between your third and fourth fingers as a "lock" (figure 8.5). This will help prevent the jesses from slipping through your fingers. Provide your bird with enough slack in the jesses that it can lift its foot if it wants. If you hold the jesses too tight, your bird may constantly fight by trying to pull its legs up, and this won't help the training process (figure 8.6). Don't let the leash dangle. If your bird attempts to fly, its wings could — and probably would — get tangled in the leash. Some custom-made gloves come with a ring for tying the leash or clipping a swivel. This can certainly be a nice feature, offering more security in case you accidentally lose your grip.

8.6b Handling a Bate

A bate is a bird's attempt to fly when it is restrained by its jesses or leash. You'll find that your bird might bate off the glove frequently during the early stages of training, especially if the traditional training approach is used. It might automatically return to the glove (figure 8.7), flap continually until exhausted, or just hang upside down. Some birds do what is called a "dead man's" bate, where they refuse to perch on the glove, ball their feet, and just drop off (a type of bird bungee jumping?). Here are a few suggestions on how to handle a bate:

- When your bird bates, make sure that it doesn't hit its wings on the floor or any other solid objects, including yourself. Move if necessary.

- When your bird bates, keep your arm steady to provide a stable perch for it to return to. Don't lower your arm. Lowering your arm brings your bird closer to the ground and stimulates it to flap more vigorously (it wants to go to the ground and you reinforce the flapping by bringing it closer).

- When your bird is bating, do not attempt to swing it back onto your glove by sharply rotating your arm or moving it up and down.

- If your bird bates repeatedly, tighten up its jesses so that if it tries to bate again it can only jump up a little and not away from you. That way it won't be able to leave your glove.

Fig. 8.7
Bate and recovery of a red-tailed hawk.
(Gail Buhl)

- Following a bate, if your bird flaps continuously and doesn't attempt to return to the glove (larger falcons and accipiters are good at this), tighten up the jesses so the bird feels the glove with its feet, place your hand on its back, and gently turn and lift it back on to the gloved hand (figure 8.8). Sometimes, turning the hand that holds the jesses will encourage the bird to turn around and come back up.

(a)

- Birds that have severe debilitating wing abnormalities (such as partial or full amputees and birds with wing nerve damage) won't be able to return to the fist after a bate and will need assistance. To help, place your hand on the bird's back and gently turn and lift it back on to the gloved hand.

- If your bird is into "bungee jumping," you might need to turn it, lift it to your fist, and keep your hand at its chest to prevent it from taking another plunge. If it refuses to open its feet, move your gloved hand slightly to disrupt its balance. Usually this will encourage it to open its feet and grip the glove.

(b)

- Especially stubborn birds can be trained to the glove by handling them initially in the dark. Sit with the bird on your gloved hand in a dark room and let it get used to the feel of your closeness and voice before gradually exposing it to better-lit conditions. This often eases the training process.

Learning how to handle your bird during bating episodes will take experience. Try these suggestions, but be patient. In time, you'll understand your bird so well that you'll be able to predict its behavior and the most effective way to deal with it. The more you handle your bird, the better you'll become at predicting a bate and averting it by slight movements of your body (slightly rotating your arm to get the bird off balance, lifting your arm so the bird is higher, etc.), eliminating the view of something threatening, and so on.

(c)

8.7 TRAINING THE RAPTOR

The rest of this chapter will walk you through training a raptor to do some basic behaviors such as stepping on the gloved hand from a tethered position, stepping back onto a perch to be tethered, getting on and off a scale, staying on the glove through doorways and confined spaces, preventing bates while on the glove, and crate

(d)

Fig. 8.9
Proper technique for fisting a raptor
from a tethered position.
(Ron Winch)

(a)
Approach bird slowly and grasp its
jesses

(b)
Untie the leash with ungloved hand
and secure a solid leash hold.

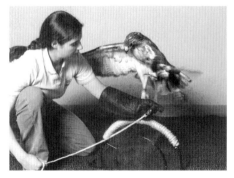

(c)
Gently press against bird's legs with
upward rolling motion to encourage it to
step up onto your gloved hand.

training. Then, the last training stage, how to introduce your birds to groups of people, will be addressed. As mentioned earlier, there are several different techniques used to train a raptor and one approach will be described here. TRC strongly suggests getting the assistance of an experienced trainer to assist you in your training adventure.

During the early stages of training, TRC recommends that your bird be tethered to a perch or hutch (see chapter 4, Housing). This will keep your bird a little calmer, and will provide it with the clear message that it cannot "escape" from you. The first question in training then becomes, "How do I encourage the bird to positively step up onto my gloved hand?"

8.7a Getting a Bird onto the Glove

Training a raptor to step up onto the gloved hand requires time and patience. Keep in mind that the bird is probably scared of you, and does not understand the action you desire or the consequences of the action. There are two different approaches commonly used to get your bird on the glove.

For both approaches, it is highly recommended to spend some time with your bird to get it used to you before trying to get it on the glove. Borrow a good book and sit with your bird daily for a while so it is not alarmed by your presence (you can even read the book out loud if you want to get the bird accustomed to your voice). Keep a little distance so you are not as threatening. If your bird appears comfortable, you can move a little closer. Your bird's "safe zone" will decrease as it becomes more comfortable with you.

1. One approach to "fisting" a bird (getting it to step up onto your gloved hand) involves just spending time, moving slowly, and letting your bird know what you want. Food is not used as a motivator.

After your bird is used to your presence in its mew, slowly approach it, grasp its jesses (figure 8.9a) between the thumb and index finger of your gloved hand, and untie the knot with your other hand (figure 8.9b). Retain hold of the leash and encourage your bird to step up by gently pressing your gloved hand into its legs with a slight upward rolling motion. If your bird refuses to budge, slowly pick up its middle toe (on the foot farther from you) and place it partly on your glove. Then, with an upward rolling motion, gently lift the bird slightly. It should place its other foot on your glove and reposition itself (figure 8.9c,d, e).

Once the bird is on your glove, put the jesses in the lock position (between your third and forth fingers), loop the leash around your fingers, and remove the perch from your bird's view (turn your body, exit the area). If you stand near the perch, your bird will undoubtedly bate for it.

The most difficult aspect to overcome with this approach is that if the bird is not an imprint, there is no motivation for the bird to want to be close to you and step up. You are forcing the bird to get on the glove (you are not giving it any choice). When fisting birds in this fashion, raptors will demonstrate their "fight" or "flight" response to the stressor, which is you. In general, many owl species tend to adapt to this style of training better than hawks as their main goals are to stay sedentary and inconspicuous during the day. They tend to stand their ground and "fight" by exhibiting defensive postures to try to dissuade an intruder. Most hawks, on the other hand, have a "flight" strategy and during early training sessions will probably bate away from you repeatedly as you approach.

(d)
Slowly lift bird to encourage it to place both feet on glove.

2. Another approach to fisting involves operant conditioning, with food as the motivator.

In the first several days, after you sit with your bird and catch up on your reading, leave your new educator its food (a half meal) by letting it see the food, slowly approaching it, and carefully placing it on the perch with your gloved hand. If your bird bates, wait until it returns to the perch to advance. You don't want to teach your bird that if it bates, food will appear on the perch. It must be perched and calm before food will be left. If the bird is extremely anxious, fast it for a day or two (if the species is small, either fast it only one day and/or leave it a small amount of food). Don't reward bad behavior.

(e)
Left bird away from perch.

Behavior	Cue	Conditioned Stimulus	Positive Reinforcement
stepping to glove ⟶	gloved hand at foot level ⟶	click/whistle ⟶	food tidbit

Your bird will quickly make the association between you and dinner and very soon it will sit on its perch while you leave food. Food is the positive reinforcer. That is why it is suggested to only provide half of a normal sized meal until the bird calms down a bit (this should be only a few days). If the bird isn't hungry at all at this stage, there will be no motivation to stay on the perch. Also, once the bird remains perched for you, stay in the mew while it eats. In the wild, raptors are most vulnerable when they are eating. Let your bird know that it will be safe if it eats in front of you. This will greatly improve your relationship.

When the bird is calm on the perch, encourage it to take its meal out of your glove with its beak. Don't offer it tidbits at this stage; offer it the total day's allotment. The food needs to look enticing to help the bird overcome its fear of the glove. If it won't take it, don't leave any food (unless the species is small, then leave it half a normal meal size). The rules have changed. If the bird wants to eat, it now needs to take food from your glove.

Once it is taking food out of your glove for a few days, you are ready to encourage the bird to place a foot or two on the glove to eat. Hold the meal in your gloved hand such that a nice portion is visible but you have a secure hold on the rest so the bird can't grab it entirely from you. As the bird tries to take it, it will realize that it must bite off pieces and will quickly learn to step up so it can hold and tear the food more easily. As it does so, make a bridging sound (click or whistle) once. After it has eaten on your fist like this for a few days, you are ready to stand with your bird in the mew.

As the bird is eating, slowly move the jesses with your bare hand so they are placed between the thumb and index finger of your gloved hand. Watch the bird carefully as you do this. If it looks at your hand and/or loses focus on the food, stop. It might be ready to foot or bite your hand. If things go smoothly, put the jesses in the lock position, untie the leash with your bare hand, wrap the leash in your fingers and slowly stand. When the bird is almost finished eating, bend down, retie the leash, and ask it to calmly get back onto its perch. Gently press the birds back legs to the perch. It should step back. When it does, make the bridging sound and let it take any remaining food with it. Then slowly retreat. It may only be on the glove for a minute or two when you are standing, but you set your bird up to succeed.

This whole encounter is a very positive experience. You will soon find that the bird will make smooth transitions to and from your glove when you make the sound, whether or not you have food. Eventually, the cue of the glove will be sufficient to elicit the behavior of stepping up. However, to keep the association solid, it is a good idea to reinforce the behavior occasionally with a food reward. For this intermittent reinforcement, keep the food hidden in your glove until the bird steps up. Then immediately let the bird see and take the reward.

Even though food is a motivator in this style of training, a bird's weight does not need to be dropped excessively to get it to step onto the glove for its meal. It may end up missing a meal or two (or get a smaller meal) but after that, if not pushed too fast, it should respond rapidly.

The risk in recommending food as a motivator is that someone may interpret the procedure to mean that a bird should be fasted indefinitely until it responds as desired. NO! As mentioned above, a large bird can be fasted a day or two or given smaller-sized meals to increase its motivation for food. A small species (American kestrel, merlin, burrowing owl, Northern saw-whet owl, sharp-shinned hawk, etc) should not be fasted more than one day and it is highly recommended to just give smaller amounts of food instead to increase the motivation for food.

If you try operant conditioning to train your bird to step onto the glove and use food as a positive reinforcer, you should see at least a small hint of improvement within a few days. This may be something as minor as the fact that the bird doesn't bate away from you until you are within three feet (0.9 m) versus six feet (1.8 m) when you started. Be patient. Reward small successes and remember the concept of baby steps. You don't want to push the issue to the point where it threatens your bird's health. As you will experience, the first week is really a critical adjustment phase (for both you and your bird); once you move past it, training does get easier.

8.7b Returning a Bird to a Tethered Position

Approaching the perch
When you move toward the desired perch with your bird, there's a good probability that the bird will bate for it in anticipation. To avoid this, turn the back of the bird to the perch as you approach it. You can distract your bird from turning and looking at the perch by slowly moving your free hand, or by providing your bird with a tasty treat.

Tying the knot
With the bird held securely on your glove by its jesses, tie the bird's leash to the perch using a falconry knot (figure 4.12). This knot is easy to do with one hand, is secure, and is easily removed with one hand. With a little practice, you can become an expert knot maker in no time. Do not let the bird jump to the perch until it is securely tied. If you do not have complete control of the bird before it is tied, a couple things could happen: the bird's leash could slip through your fingers and the bird escape, or the bird could jump/fly at you when you are tying the knot, both undesirable behaviors. Also, when tying the knot do not hold the bird such that its feet are near your face. It only takes a split second for injuries to happen.

Transferring the bird to its perch
Once the leash is tied securely to the perch, you can use one of two techniques to transfer the bird to its perch. You can turn your bird forward so it can hop off the fist, which usually needs little encouragement. If your bird is reluctant, hold it slightly lower than the

perch. Encourage it to step back onto the perch by gently pressing the back of its legs against the perch.

8.7c Weighing a Bird

Weighing your bird should be a consistent part of your handling routine. Place your scale on a table with no objects nearby that your bird could hit if it bates or flaps its wings. TRC recommends training your bird to back onto a scale instead of having it jump forward to the scale. With the latter, once your bird is trained it probably will bate for the scale before you reach it, anticipating the positive reinforcement for the behavior (which could either be a tidbit of food or just the mere act of getting off your glove). This will be less likely to happen if you train the bird to back onto the scale.

Fig. 8.10
Weighing a trained red-tailed hawk.
(Ron Winch)

Once again, there are two basic approaches to training this behavior. They are similar, except that one rewards the behavior with food and the other does not. The basic idea is to approach the scale, remove the loops of leash from around your fingers, back your bird up to and slightly below the perch and ask it to step back. If you touch the bird's legs to the perch, it should step back onto it. If not, you may need to carefully lift one of its feet and place it on the perch and then lower your gloved hand to encourage the bird to place its other foot on the scale.

Behavior	Cue	Conditioned Stimulus	Positive Reinforcement
stepping on & off scale ⟶	pressing legs against perch ⟶	whistle or click ⟶	food tidbit

Don't let go of the jesses. Instead, hold the jesses and swivel up even with the bird's feet (figure 8.10) — doing this consistently will eliminate variances in weight due to the equipment. Quickly read the weight and then encourage your bird back onto your glove by gently pressing into its legs with an upward rolling motion. Eventually, this process will take little effort, and the bird will readily step back onto the scale and then up onto your glove.

With diurnal birds, you can easily use a food reward to train smooth transitions to and from the scale. When you press your bird's legs against the scale and it begins to step back, whistle or click and be prepared to offer a food tidbit when the bird is on the scale. However, do not offer the food with your bare hand — use a small forceps or stick; otherwise, your bird may foot your hand in anticipation of the food. In the same fashion, you can train the bird to get back on the glove. Press gently into the bird's legs and click. When it is on the glove offer a food tidbit. Soon, you will just have to whistle or click to get your bird on and off the scale. Intermittently rewarding the behavior with a tidbit will keep it dependable.

8.7d Doorway Passage

Once your bird accepts your gloved hand and you are able to walk it around, you'll notice that close spaces such as doorways, stairways, elevators, and hallways are "scary" and usually stimulate your bird to bate. Repeated exposure to these situations will help minimize the bird's response. Also, distracting the bird with food or hand movements as you pass through halls or doorways will help divert its attention. Before your bird is aware, it will have passed through the doorway without even realizing it.

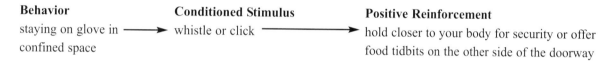

Behavior   **Conditioned Stimulus**   **Positive Reinforcement**
staying on glove in ——→ whistle or click ——————————→ hold closer to your body for security or offer
confined space   food tidbits on the other side of the doorway

You can also use operant conditioning to reward your bird for staying on the glove while going through these spaces. If your bird stays on the glove, provide your bridge immediately, and then follow it up with positive reinforcement. If the bird bates, do not provide the reinforcement. You can practice this in short mini sessions, where you go through doorways several times and reward only those times that the bird stays perched.

8.7e Crate Training

Before your bird will be ready for vehicle travel it must feel comfortable in its travel carrier. Guidelines for choosing a transport container are outlined in chapter 9, Transporting Raptors. Here, training your bird to smoothly enter and leave the crate will be addressed.

To begin, place the travel crate up off the floor. Your bird will be less reluctant to enter the crate if it is higher. Back the bird into the crate and press its legs against the perch (a process similar to encouraging it to step onto a scale). The bird should be accustomed to stepping back at this point and might back onto the perch quite nicely. Smoothly remove your hand and close the door.

Owls are usually pretty good about going into a crate. Hawks and eagles, on the other hand, are a little more nervous about the experience and may repeatedly bate before entering the crate. If this happens a few times, there are several things you can do. When backing your bird into its crate, keep your bird close to your body and use your body to block any visual or physical access your bird might have to an "escape" route. Move slowly and smoothly until your bird is in the crate. If you are still having difficulty, place your bird in its crate in the dark and leave it there for a few minutes. Then, turn on a light and remove the bird. Repeat this process until the bird will go smoothly into the crate with the lights on.

If, despite your attempts, your bird is still terrified of entering a crate backwards and repeatedly jumps toward you, train the bird to enter forward instead (as described below). Once a bird is comfortable with the crate in general, you can switch to back loading if you prefer. Sometimes, birds trained to front load bate in anticipation when seeing the crate. Overzealous entries can lead to wing and feather damage.

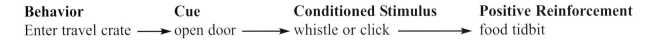

Behavior	Cue	Conditioned Stimulus	Positive Reinforcement
Enter travel crate ⟶	open door ⟶	whistle or click ⟶	food tidbit

To train your bird to front load, place a small piece of food in the doorway and place your bird so it can reach it. Then place tidbits progressively farther into the crate, but hold your bird at the crate door. If it really wants the reward, it will have to get off your glove either onto a perch in the crate or the bottom of the crate (if no perch is provided). Let your bird eat the tidbit, and then present your glove to provide your bird a safe place to return to.

Once the bird is going into the crate regularly, you can start shutting the door for increasing periods of time. Eventually, you'll no longer need food and the bird will go willingly into the crate. Once again, occasional reinforcement will help keep this behavior solid. The interesting thing about crates is that they provide a safe, dark place for birds, away from people and other stressors. This in itself is often positive reinforcement. When a bird learns that it gets quiet time in a crate, it usually goes in willingly.

To remove your bird from its crate, slowly open the door, grasp the jesses, and encourage the bird to step up onto your glove. Cover the opening of the carrier with your body so the bird doesn't fly out. If you have developed a conditioned stimulus (whistle or click), use it to help cue the bird that you want it to step up.

Once your bird has spent time in the crate, you can gradually begin moving the crate, loading it into a vehicle, and then traveling. Initially, take your bird for short trips to familiarize it with moving in a vehicle. That way, when you take it to its first off-site program it won't be so stressed from the travel that it's too nervous and edgy to sit calmly on your glove or a perch (chapter 9, Transporting Raptors).

8.7f Stage Fright: Introducing the Bird to People

Program birds

Now that you've trained your bird to accept you, your handling, and automobile travel (if you will be taking your bird off-site), you are ready to introduce it to the world of the public program. Start the bird with small, quiet groups of people and watch how it handles the situation. Remember, signs of stress such as constant bating, panting, or rapid head turning to find an escape route all indicate that your bird isn't comfortable. Respect these actions and limit the time your bird is confronted with the situation.

If you have more than one education bird and want to show them to a group at the same time, be aware of where the birds are placed relative to each other and how that might affect them. Birds that are natural predators or prey of each other should not be placed close to each other (for example, barred owls and screech owls, great horned owls and peregrine falcons, American kestrels and common barn owls). Otherwise they may be edgy and defensive. Either keep them a good distance apart (so one won't bate toward the other) or provide a visual barrier. This same principle should be applied to their housing, as described in chapter 4.

With time and exposure, you'll find that your bird will become more accepting of groups of people (as long as they're kept at a respectable distance) and close activity. Don't work your bird too hard. Provide it with "days off." Like humans, birds need a break to recuperate and relax.

Display birds

Up to this point, we've said little regarding birds that aren't routinely handled for on- or off-site programs and are kept for display only. Some people don't handle their display birds except for physical exams or relocating them. Keep in mind that the more a bird is exposed to something, the more accepting and less stressed it will be when faced with it. Manning a display bird and handling it frequently, even for short periods of time, will increase its comfort level with people and create less stress when the public views it.

Whether or not you choose to handle your display bird regularly, you must make sure that it feels comfortable and safe in its display environment before introducing it to groups of people (just as your bird must accept you and your glove before you take it on a program). Chapter 4 (Housing) provides guidelines for creating such a safe haven. If your bird is hanging on the walls of the enclosure, constantly flying away from you as you approach, or frantically walking back and forth along the perimeter of the mew looking for a way out, it is not comfortable and not ready for people. You might need to tether it for a while to inhibit such activities (see chapter 4, Housing). To help your bird adjust to its new surroundings, it is important to provide it with some consistency. Follow a regular cleaning/feeding routine (person involved, time of day, etc.) and try to stay out of the bird's "safe zone."

Once your bird accepts its new home, you can begin moving around it and performing some activities. At first, the bird will probably watch you or hide in its shelter box. You can slowly approach the mews, but don't stare at your bird as you do so — that would seem threatening and make the bird uneasy. Out of the corner of your eye, observe the bird's behavior; it will tell you how it's feeling about your proximity. Remember, repetition will increase your bird's comfort level with a situation. The more you are in its view while doing something other than focusing your attention on it, the more comfortable it will be in your presence.

Once your bird shows little or no reaction to your presence, you can bring a few people to observe it. Make sure that there's a barrier (such as a rope or fence) that prevents people from either invading the bird's personal space or getting hurt by putting fingers through the mesh, etc. If your bird handles this situation well, you can gradually increase the number of people.

8.8 COMMON TRAINING MISHAPS

Over the years, TRC has consulted handlers on numerous raptor behavior "problems." In many cases the undesirable behaviors developed because they were unknowingly reinforced by the trainer's actions. Using some of the training tools shared in this chapter, these behaviors could be replaced with more desirable ones, creating a positive relationship for both the bird and handler. Let's take a peek at a few of these situations:

1. A bird bates a few times and the trainer immediately tethers it to a perch or returns it to its mew. The trainer is training the bird that in order to go back to the security of its mew, it needs to bate. A bird should not be returned to its mew until it is calm and has sat quietly on the glove for a few minutes. This is the desirable behavior that should be rewarded. If a bird is highly stressed, it should be removed from the scene. However, it should still be

asked to calm down on the glove in a quiet area before returning to its housing.

2. A bird has at least partial flight ability and the trainer thinks it should get exercise. So, the bird is attached to a creance (long line) and tossed into the air to fly several times a week. First, to keep healthy, a bird does not "need" to fly. Second, by encouraging it to fly off a gloved hand in this fashion, it is being trained to bate. If the bird is being trained to fly for demonstration purposes, the situation is different because operant conditioning techniques are used. The trainer provides a specific cue when it is appropriate for the bird to leave the glove, and the bird is trained to reach a certain destination for a reward.

3. A new program bird is free-lofted and won't get on the handler's glove. So, the handler chases it, grabs it by the legs, and puts it on the glove. He or she then leaves the mew and expects the bird to sit contently on the glove during a program. The bird will no doubt be stressed, as the act of being chased and grabbed is very threatening — the handler is essentially a predator. What is the motivation for the bird to allow the handler to approach, or for it to step up and stay on the glove? This bird should either remain a display-only bird, or be put through a formal training process. Tethering it during the training (and depending on the species and individual, potentially always) will help develop a clear communication path and begin the training process. The bird should be taken off program use until it can be handled in a positive manner.

4. A handler wears gloves on both hands when training or handling a program raptor. Most permanently disabled birds have been negatively handled with two gloves during the rehabilitation process. They associate two gloves with being grabbed and will be edgy, defensive, and often bate excessively with a second glove in sight. Training and handling a program raptor should be positive and a single training glove, visually different from a grabbing glove, should be used. This will send a clear message to your bird that it will not be grabbed when handled on the training glove. People who have used two gloves did so because they feared getting bit or footed. If you are afraid of your bird, you should not be handling or training it.

5. A raptor bates when its mew door is open and the handler releases the bird and allows it to fly in. Raptors are intelligent birds and when routines are developed, they will anticipate a specific behavior and often act before asked to. Bating towards the mew, travel carrier, or scale are all undesirable behaviors that will only be reinforced if the bird is allowed to reach its destination following the bate. A raptor should perch on the gloved hand until asked to step off into the mew, onto the scale, or into its travel carrier. This will prevent the bird from becoming injured in its overzealous attempts to reach its destination and will keep you in charge of your bird's behavior.

8.9 TOP TRAINING TIPS

Well, you've just about made it through this chapter and hopefully you have a better understanding of how to communicate with your bird. To help summarize, here are some top training tips:

1. Get prepared ahead of time. Get a mentor; go to training workshops, read animal training techniques.

2. Set specific realistic goals.

3. Break down each goal into baby steps.

4. Set your bird up to succeed; do not ask it to exceed its capabilities.

5. Never try to force your bird to do something it is not ready to do.

6. Be consistent, repetitious, and develop routines your bird can count on.

7. Don't move on to another behavior until the previous one has been successfully performed on a consistent basis.

8. Time positive reinforcement carefully to encourage the correct behavior.

9. Do not use forms of punishment to "reprimand" or eliminate undesirable behaviors.

10. Several short training sessions throughout a day can be

more effective than one or two longer sessions (human and bird attention spans can be short).

11. Stop a session if your bird is too stressed. You may need to back up a step and take more time.

8.10 SUMMARY

As you can probably tell, there's so much to learn about bird behavior and basic training techniques that volumes could be written on the subject. The information presented here is intended to provide a foundation upon which you can build your skills in bird training. You might find that some of the techniques presented here work very well for your bird, while others do not. Every bird is different, and training techniques need to be molded to your specific bird as you build a relationship with it.

TRC strongly encourages you to prepare yourself before you start handling a raptor; read some of the materials suggested here and contact experienced bird handlers for tips. The more information you have going into the training process, the more successful you will be.

8.11 SUGGESTED READINGS

Arizona Falconer's Association. 1990. *A Manual for the Apprentice Falconer.* Phoenix.

Beebe, F. L. 1992. *A Falconry Manual.* Surrey, Canada: Hancock House.

Beebe, F.L., and H. M. Webster. 1994. *North American Falconry and Hunting Hawks.* Denver, CO: North American Falconry and Hunting Hawks.

Heidenreich, B. 2001. *Getting in Touch with Their Feelings: Developing Sensitivity to Bird Behavior.* www.naturalencounters.com.

International Association of Avian Trainers and Educators Newsletter. IAATE, c/o Minnesota Zoo, 13000 Zoo Blvd., Apple Valley, MN 55124, 612-431-9356.

Malina, C. *Interpreting and Influencing Animal Behavior: An Essential Keeper's Tool.* 1999. www.naturalencounters.com.

Martin, S. and S. G. Friedman. *Training Animals: The Art of Science.* Presented at the ABMA conference in Baltimore, Maryland, 2004. www.naturalencounters.com.

Martin, S. *What's In It For Me?* Presented at the International Associations of Avian Trainers and Educators Conference in Toronto, Canada, 2004. www.naturalencounters.com.

Martin, S. *The Anatomy of Parrot Behavior.* Presented at International Association of Avian Trainers Conference, Monterey, California, August 2002. www.naturalencounters.com.

Martin, S. *Enrichment: What Is It and Why Should You Want It?* Presented at the World Zoo Conference, Pretoria, South Africa, October 1999. www.naturalencounters.com.

Martin, S. *Public Animal Programs and Their Role as Affective Forms of Conservation Education.* 1998. www.naturalencounters.com.

Martin, S. *Training: A Critical Component of Enrichment Programs.* 1997. www.naturalencounters.com.

Oakes, W. 1993. *The Falconer's Apprentice: A Guide to Training the Red-tailed Hawk.*

Warrens, WI: EagleWing Publishing.

Parry-Jones, J. 1994. *Training Birds of Prey.* Devon, England: David and Charles, Brunel House.

Pryor, K. 1999. *Don't Shoot the Dog! The New Art of Teaching and Training.* Toronto and New York: Bantam Books.

8.12 SUGGESTED VIDEOS

Martin, S. 1994. The Positive Approach to Parrots as Pets Series. Tape 1: Understanding Bird Behavior. Winter Haven, FL: Natural Encounters, Inc.

Martin, S. 1994. The Positive Approach to Parrots as Pets Series. Tape 2: Training through Positive Reinforcement. Winter Haven, FL: Natural Encounters, Inc.

Chapter 9:
TRANSPORTING RAPTORS

You may come across many instances when you are involved in transporting a raptor. The first time may be when you are on the receiving end, acquiring a new education or display bird from another facility. After that, you may play the role of chauffeur, escorting your bird to educational events, to your veterinarian for medical care, or maybe to the airport so it can fly the friendly skies. You undoubtedly will want to make the ride as comfortable as possible and meet the special needs of your passenger.

Safety and comfort are the two major considerations when transporting a raptor. The Raptor Center has developed transport protocols that address these issues for both trained raptors and "wild," untrained raptors (display-only birds, or birds first entering an educational program) traveling via ground or air.

When you're transporting your raptor to schools or other educational events, keep in mind that the smoothness and comfort of the trip will greatly affect your bird's disposition during the program. A rollercoaster ride can produce an edgy, anxious bird whose usefulness in the program will be diminished. To provide your raptor with a calm and comfortable vehicle ride, it's important to choose an appropriate carrier and pay attention to the special needs of your unusual passenger.

9.1 CARRIER FEATURES

The first step in making your bird comfortable and safe during car or plane transport is to provide it with an adequate carrier. When choosing the best carrier, there are several things to consider.

9.1a Size

A transport carrier should be large enough so your bird doesn't touch any part of the crate when its wings are closed, that it can make positional adjustments to make itself comfortable, that it can perch with sufficient head and tail clearance (if a perch is desired, it should be one inch higher than the tail length), that it can load and unload comfortably, and it can mute (go to the bathroom) without damaging wings or feathers. It should not be so large, however, that the bird is encouraged to move excessively and jump around, increasing the potential for injury. Recommended sizes of several carrier types are provided in tables 9.1 and 9.2.

9.1b Strength

Carriers can be made out of a variety of materials, including sturdy plastics and plywood. It's your job to ensure that whatever material is chosen is strong enough to contain your raptor safely, and light enough to allow the handler to move it as necessary.

9.1c Padding

The Raptor Center highly recommends padding the inside of the transport carrier. Non-Berber (unlooped) carpeting glued to the inside walls and ceiling work well. Carpet foam covered with plastic sheeting will also work, but keep an eye on it. Some birds get "bored" and pick at the plastic and eat pieces of foam. Avoid securing the padding with duct tape as even this "miracle" tape can lose its adhesiveness during travel and the pads can jar loose.

9.1d Perch

TRC does not recommend providing a perch when transporting an untrained raptor. These birds are often nervous, not accustomed to a confined space, and tend to move around a lot.

For trained raptors, on the other hand, there are pros and cons to providing your bird with a perch during travel. On the positive side, most raptors are more at ease, and their feathers (especially tail feathers) are less likely to break, if they are perched on something above ground level. Also, trained raptors can be more easily loaded and unloaded from a carrier with a perch.

In a carrier, a perch must be securely attached so it doesn't slide around. Freestanding perches should not be used during transport. For plastic or wood carriers, a sturdy perch can be attached using stove bolts passed through holes drilled into the side of the carrier and then lengthwise

Fig. 9.1
Back wall perch in crate.
(Ron Winch)

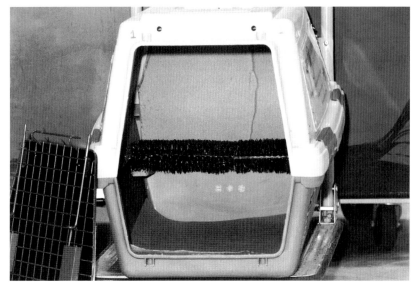

into the perch. A 2-inch x 6-inch (5.1 cm x15.2 cm) piece of wood wrapped with three-strand manila rope or AstroTurf is good for most species. The bird should perch on the 2-inch (5.1 cm) side and the 6-inch (15.2 cm) should be positioned down towards the floor. This perch reduces the possibility that the bird can get caught underneath it. A rectangular shelf perch covered with AstroTurf and attached to a side or back wall works well for falcons and can be modified for hawks and owls as well (figure 9.1).

On the negative side, a perch will increase the size of carrier needed as the bird will need greater head and tail room. This will make the crate heavier, and if you have multiple birds, you may need a larger vehicle to transport them in. Also, keep in mind that for airline travel, the handler loses control of the crate once it is checked in and fast jerky movements could cause the bird to fall off the perch and potentially injure itself.

If you decide to transport your bird in a carrier without a perch and will be traveling for a few days, you will need to bring a portable perch to tether your bird to in order to provide comfort and prevent foot sores. Relatively few species naturally perch on flat surfaces and forcing other species to do so for longer lengths of time can lead to foot problems.

9.1e Flooring

The covering of the floor should be easy to clean. For birds that perch well during travel, newspaper or thick kraft paper can be placed on the bottom to collect mutes. For birds that are transported without a perch, Monsanto AstroTurf, stadium matting, padded indoor/outdoor carpeting, or non-looped carpet squares work well. They provide your bird with a comfortable resting surface that helps absorb mutes (matting and carpeting) or keeps mutes away from your bird's feet (AstroTurf). These coverings should be securely attached to the bottom of the carrier so they don't slide around during transport.

9.1f Door Position

Something many people forget when designing or choosing a carrier is the position of the door. If you're transporting a trained raptor by car, the carrier should have a door on the front, not on the top. There are two reasons for this.

First, it's very difficult to remove and replace a trained raptor through the top of a carrier. Raptors like to be high and will not cooperate when being lowered into a carrier (they'll probably resist by trying to walk up your arm and flapping their wings).

Second, approaching your bird from above is very threatening (predators and food competitors often approach from above) and will cause distress. However, as you will see, for an untrained raptor, TRC uses a darkened rectangular carrier with the opening on the top. It is much easier to remove a "wild" bird from a larger top opening than through a relatively small front opening.

9.1g Light Penetration

One very important consideration is the amount of light that penetrates into the carrier. Light stimulates activity and might make your bird restless during transport. Make sure, therefore, that the inside of the carrier is as dark as possible while still providing adequate ventilation. Ventilation holes or slots should be placed on the lower part of the carrier walls (to prevent the bird from seeing out, and people from looking in), and tiny holes for cross-ventilation should be placed near the top of the walls. Place holes on opposite sides of the carrier to promote cross-ventilation.

9.1h Ease of Cleaning

Fig. 9.2
A Vari-kennel modified to transport an educational bald eagle.

One last thought regarding the bird's carrier is to make sure that it lends itself to easy and frequent cleaning. It's essential to keep your bird's housing structure, and transport carrier clean and disinfected to prevent disease (see chapter 4, Housing).

9.2 CARRIER OPTIONS

Keeping in mind the carrier design features mentioned above, there are several types of carriers that can be modified to provide your bird with safety and comfort during transport.

9.2a Automobile Travel

A few different carrier types are recommended for automobile travel:

Vari-kennel
Vari-kennels (figure 9.2) are appropriately sized dog and cat plastic carriers that can work very well for a variety of raptor species (table 9.1). However, birds that are nervous travelers, or have more hyperactive personalities don't travel well in these carriers because they can see through the door and windows. You can cover these openings with burlap, pegboard, shade cloth, or corrugated plastic, but you must take great care not to compromise ventilation. Removable solid window covers can be effective and allow the crate to be ver-

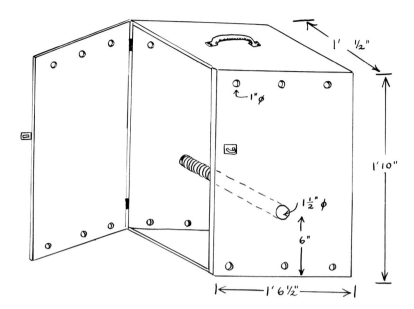

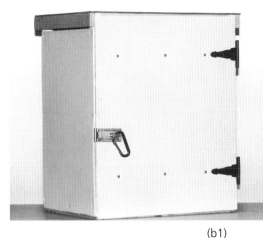

Fig. 9.3
Giant hood.

(a)
Diagram of a typical giant hood
constructed of plywood.

(b1)
A giant hood constructed with a wood
frame and corrugated plastic sides.
(Ron Winch)

(b2)

(c)
A plywood giant hood constructed
with a side perch.

satile during transport. If a Vari-kennel is your carrier choice, TRC recommends models that allow the door to be open from either the left or right side. This offers more flexibility. Avoid models with guillotine-style doors as it can be difficult to maneuver a bird in and out of this arrangement and there is the possibility that the door will actually fall down and hit your bird during loading or unloading.

Giant hood

An alternative carrier can be constructed out of plywood, or sturdy plastic, such as corrugated plastic (figure 9.3). This type of carrier is quite dark, so it tends to keep birds calmer. However, it also retains heat, so it must be constructed carefully to provide adequate ventilation and cooling (table 9.2).

9.2b Air Travel

TRC has safely transported a wide variety of species via airplane, using both private and commercial carriers. With an appropriate transport container, airline travel can be a safe and reliable means of transport. If you need to ship your trained bird via the friendly skies, you can use Vari-kennels or giant hoods, with some of the modifications listed above.

First, cover all doors and windows, as explained earlier, to reduce stimulation from the airport/cargo areas (and reduce the temptation for people to peek in) — but don't compromise ventilation. If the carrier has a perch, it must be securely fastened to the carrier so it won't move during transport. TRC also recommends securing the door with plastic tie wraps; these deter curious people from opening the door and peaking in. However, due to airport

Please position this box so the actual aircraft matches this illustration.

If there are any problems with this shipment please call:

The Raptor Center Main #
    612-624-4745

Enclosed are the Importation Permit & Health Certificate for one:

To Be Received By:

Fig. 9.4
Envelope to apply to an airplane travel crate indicating proper position of the crate relative to the plane.

Fig. 9.5
Rectangular carrier used to transport untrained raptors on airplanes.
(Gail Buhl)

security measures, it is wise to secure the door after it has been accepted at the airport (see 9.5, Travel Tips).

If you provide your bird with a perch during airline travel, it is important to notify the airline how to properly position the crate such that it is perpendicular to the line of motion. This way the bird will move side to side instead of tip forward and backward in response to plane movements. One way to do this is to apply a label or envelope on the crate showing the position of the crate in relationship to the airplane (figure 9.4).

If you are transporting an untrained bird, TRC recommends using an appropriately sized rectangular plywood carrier (figure 9.5, table 9.2) or a well-padded, darkened Vari-kennel. Darkened carriers with limited visual access to the outside will calm the bird and make it more at ease as it travels.

9.3 CRATE TRAINING

It is important to realize that most trained raptors need to become familiar with a transport carrier, learn to accept it, and make smooth entrances and exits. Work with your bird to get it used to the carrier before you start traveling (see chapter 8, Training).

9.4 TRAVEL HAZARDS TO AVOID

When your bird travels, not only are you often the chauffer, but also your bird's travel agent, arranging for the most comfortable accommodations possible. Your bird does not have the ability to refuse inadequate housing or complain about poor service. Therefore, it is up to you to be mindful of your passenger's needs and create the best environment possible.

1. Never house your bird or travel with it in a wire cage. Wire can cause serious injuries to your raptor. If the bird jumps onto the wire, for example, it can lacerate its toes and damage tendons. Abrasions can easily result if the bird rubs any part of its body against the wire. Toes and other body parts can get caught in the wire, and feathers can be severely damaged.

2. Don't use freestanding perches in a carrier during transport. They will often tip over, resulting in potential injury to your bird.

3. Do not include food and water in a carrier during transport. Most birds won't eat during a trip, and water tends to spill. If birds do eat, they could regurgitate and aspirate fluid into their lungs.

4. Do not feed your bird for a minimum of four hours prior to airline transport or long car trips. If you offer your bird food prior to that time, do not offer a full meal. Like people, birds can get airsick and vomit during a flight.

5. Never leave your bird in a parked car for any length of time — even a few minutes — during warm weather. To be safe, the temperature in the vehicle should not exceed 70° F (21°C). Keep in mind that the raptors are in a crate inside the vehicle and the air is somewhat stagnant. Birds can overheat very quickly in a warm/hot vehicle without proper ventilation and cooling measures. Even in a short time, a bird can be overcome by heat, become hyperthermic, and die. Also keep in mind that different species have different heat tolerances (chapter 2, Selecting a Bird for Education).

9.5 TRAVEL TIPS

Now that you are aware of some serious things to avoid while transporting your raptor, here are a few things you can do to make your bird's transport safe and comfortable:

• When you place a carrier into the transport vehicle, position it so the bird is perpendicular to the line of motion. When the vehicle stops and starts, it's easier for a bird to balance from side to side rather than rock back and forth. (Remember having to stand during a bus ride?)

• Each bird should have a designated carrier. This is better for your bird's health and you will not have a lot of preparation if your birds have different travel needs (different perch types, covered vs. uncovered windows and doors, etc.).

• Do not house more than one bird in a carrier, even if they are normally housed together.

• Make sure to provide adequate ventilation both of the carrier and the vehicle. This is a much greater challenge on hot summer days. Some species, such as Harris's hawks, are very sensitive to exhaust and can be killed quickly if exposed to fumes.

Fig. 9.6
Applying a tail protector.

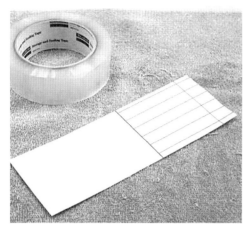

(a)
Materials commonly used to make a tail protector (stiff envelope covered with packing tape for waterproofing).

(b)
Keep the tail feathers in a natural closed position and slide them into the protector.

(c)
Keep the under tail coverts out of the protector.

• Provide a comfortable temperature for your bird during transport. TRC strongly advises transporting birds in air-conditioned vehicles when the temperature is above 75°F (23.9°C). Applying shades to the windows of your vehicle can also help to keep the temperature cool in warm weather. On the flip side, if birds are housed outdoors during cold weather, transport them in a comfortable but cool temperature (i.e. don't crank the heat on high for the trip, even if you like it toasty).

• You can transport your bird with a swivel and leash attached provided you are traveling for a relatively short period of time (an hour or less), or your bird likes to bolt out of the crate. However, do not leave a swivel and leash on your bird if it is traveling by plane, or is an anxious traveler who tends to turn around and move a lot during transport.

• If you must listen to music when transporting a bird, be considerate. Choose soft, quiet music. You can imagine how a bird would feel and "perform" after listening to loud heavy metal music on the way to a program. Sometimes, constant soft background noise will actually keep a bird calmer.

• If you are transporting an untrained bird that it not accustomed to traveling, apply a tail protector to its tail feathers to prevent damage (figure 9.6).

• All crates should be properly labeled with identifying markers such as "live animal" stickers, arrows indicating that the carrier should always remain in an upright position, and your contact information (name, address, cell phone number, phone numbers of other potential contact people). This is especially critical if your bird is traveling by air. TRC also places a sticker of a plane on the crate illustrating how the crate should be positioned so the bird's body is perpendicular to the line of motion. Make it as easy as possible for airport personnel to handle the crate properly and to get assistance if a problem arises.

• When making arrangements to transport your bird via plane, find out about the different ways the bird can travel (cargo, counter to counter, VIP). Choose the one that best suits the bird and your situation. Also, make sure you check with the airline for transport rules and restrictions. Most airlines have strict rules regarding the ambient temperature ranges at which they will transport animals, carrier size and preparation, and security measures. Many airports require that the inside of the crate be inspected prior to accepting it for shipment. This may involve removing your raptor at the airport to have a full crate inspection conducted. Finding this out ahead of time can save you great pains if you were planning on having a non-handler deliver the bird to the airport.

- Bring copies of your state and federal permits with you when traveling. This will avoid problems if you get stopped or are otherwise questioned.

- Always bring a handler's glove and a bird emergency kit (appendix E) when traveling. If traveling by plane, it is best to place these in your carry-on bag in case your luggage gets lost (always prepare for the unexpected).

9.6 SUMMARY

Raptors can be transported safely by car or plane. To ensure that your bird travels comfortably and calmly, you must choose an appropriate carrier design and modify it according to the bird's temperament and needs. Vari-kennels or plywood carriers work well for both types of travel. Special modifications such as carrier padding may be required to comply with airline codes. Be sure to check with your airline to find out what other preparations are necessary and what restrictions might be in place.

9.7 SUGGESTED READINGS

Meng, Heinz. The Giant Hood Update. 1993. *Hawk Chalk,* Vol. 31, No. 2.

Kimsey, Brian, and Jim Hodge. 1992. *Falconry Equipment,* pp. 143–159. Houston: Kimsey/Hodge Publications.

Weaver, Andy, and Patrick T. Redig. Shipping Raptors on Airlines. 1990. *Hawk Chalk,* Vol. 29, No. 1.

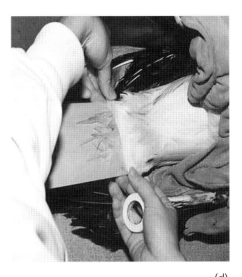

(d)
Apply a few layers of tape (Durapore (3M) or masking tape) on top of the coverts and around the protector to secure it.

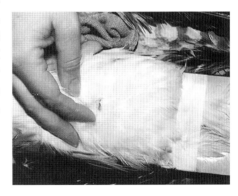

(e)
Check to make sure the bird's vent is not covered by the protector or taped feathers.

Table 9.1 Recommended sizes for Vari-kennels and giant hoods

Species	No Perch (NP) Perch (P)	Sky Kennel vendor # (when available) and (LxWxH)	Giant Hood (LxWxH)
American kestrel, boreal owl, burrowing owl, Eastern screech owl, merlin, Northern saw-whet owl, sharp-shinned hawk, Western screech owl	NP	Vari-kennel, Jr., small 19"x12"x11" (48.3cmx30.5cmx27.9cm)	12"x12"x12" (30.5cmx30.5cmx30.5cm)
	P	Vari-kennel Jr, small 19"x12"x11" (48.3cmx30.5cmx27.9cm)	12.5"x12.5"x12.5" (31.8cmx31.8cmx31.8cm)
Common barn owl, broad-winged hawk, Cooper's hawk, long-eared owl, , Mississippi kite, Northern harrier, peregrine falcon, prairie falcon, short-eared owl	NP	#100 - 21"x16"x15" (53.3cmx40.6cmx38.1cm)	16"x20"x16" (40.6cmx45.7cmx40.6cm)
	P	#200 - 27"x20"x19" (68.6cmx50.8cmx48.3cm)	18"x18"x22" (45.7cmx45.7cmx55.9cm)
Barred owl, crested caracara, ferruginous hawk, great horned owl, Harris's hawk, red-shouldered hawk, red-tailed hawk , rough-legged hawk, snowy owl , Swainson's hawk	NP	#200 - 27"x20"x19" (68.6cmX50.8cmx 48.3cm)	18"x20"x18" (45.7x50.8x45.7)
	P	#300 - 32"x22"x23" (81.3cmx55.9cmx58.4cm)	20"x20"x24" (50.8x50.8x61.0)
Great gray owl, Northern goshawk, swallow-tailed kite	NP	#200 – 27"x20"x19" (68.6cmx50.8cmx48.3cm)	18"x24"x18" (45.7x61.0x45.7)
	P	#300 – 32"x22"x23" (81.3cmx55.9cmx58.4cm)	20"x20"x24" (50.8cmx50.8cmx61.0cm)
Bald eagle, black vulture, golden eagle, osprey, turkey vulture	NP	#400 - 36"x24"x26" (91.4x61.0x66.0)	26"x30"x26" (66.0x76.2x66.0)
	P	#500 - 40"x27"x30" (101.6x68.6 x76.2)	26"x26"x30" (66.0x66.0x76.2)

Table 9.2 Recommended sizes for rectangular carriers

Species	Size in inches LxWxH (cm)
American kestrel, boreal owl, burrowing owl, Eastern screech owl, merlin, Northern saw-whet owl, sharp-shinned hawk, Western screech owl	12x8x10 (30.5x20.3x25.4)
Broad-winged hawk, common barn owl, Cooper's hawk, long-eared owl, Mississippi kite, peregrine falcon, prairie falcon, short-eared owl	20x12x 14 (50.8x30.5x35.6)
Barred owl, crested caracara, ferruginous hawk, great horned owl, Harris's hawk, Northern goshawk, Northern harrier, red-shouldered hawk, red-tailed hawk, rough-legged hawk, Swainson's hawk, swallow-tailed kite	26x14x14 (66.0x35.6x35.6)
Black vulture, great gray owl, osprey, snowy owl, turkey vulture	30x14x20 (76.2x35.6x50.8)
Bald eagle, golden eagle	34x20x22 (86.4x50.8x55.9)

Chapter 10:
RECOVERING A LOST BIRD

Whenever you keep a wild bird in captivity, there's always the possibility that sometime, somehow, it might escape. It is important for you to know that this is a real possibility and that even highly experienced handlers have lost birds both inside buildings and outside. Don't be complacent about your bird's security. Realizing that your avian educator is gone or watching it depart into the wild blue yonder is a sickening feeling. You are the bird's caretaker and are responsible for its health and safety. How can you prevent this nightmare from happening? Based on true stories, here are recommendations to prevent your bird from "making a break for it":

1. Use a vestibule (double door) system for all outdoor housing units.

2. Never leave tethered birds unattended in open areas or other areas with potential escape routes (such as rooms or mews with open windows). Not only could your bird escape, but open areas also leave your bird vulnerable to aerial and ground predators. It only takes a second for things to happen, and they do.

3. When your bird is on your glove, secure its leash by either tying it to a glove ring or "locking" it between your fingers as explained in chapter 8, Training.

4. Check your bird's equipment frequently and always before tethering in open areas. Replace it at the first sign of wear.

5. Do not tether birds with jess extenders in an open area. A jess extender only has one point of attachment between the swivel and a single piece of leather or nylon loop. If the leather or nylon extender breaks, your bird is gone. When each jess is attached to a swivel, there are two places of attachment and even if one jess breaks, the other will temporarily secure your bird.

6. Use a double falconer's knot to tether your birds in open areas (clips always have the potential to become faulty).

7. Only allow experienced people to handle your birds outside. Do not train new people on bird handling or transfer birds between handlers outside.

8. Have a safety plan for inclement weather. Straight-line winds, tornadoes, and hurricanes can all cause collapse of mews. If these are forecasted, your bird should be moved inside to a safer area (basement, cellar, solid building, house, etc. or taken out of the area entirely). Have indoor housing, such as crates, prepared to put your bird into.

Even with the most conscientious management and handling practices, accidents happen. If your bird gets loose in a building or escapes outside, don't panic. There are proven techniques you can use to retrieve it. Keep in mind that your bird will not be familiar with the surroundings, and the only destination it has in mind is a safe perching spot. Also, if your bird came from a rehabilitation facility with a healed wing or collarbone injury, it probably isn't capable of sustained flight over long distances. This should help put your mind at ease.

10.1 RETRIEVING A BIRD THAT IS LOOSE IN A BUILDING

Let's start with the easiest scenario. Your bird zips past you out of its crate or you lose your hold and it flies from your glove in an enclosed space. Immediately close all doors and windows to contain your educator. Then, there are several things you can do to regain control of your bird. The first few techniques are passive, encouraging your bird to move into a secured position.

1. If the bird is reachable, approach it slowly and ask it to step up onto your glove. If it eats off the glove, offering a tidbit may provide it with extra encouragement.

2. Place its crate nearby with the door open and back off. If your bird is crate trained, it may enter the crate for security.

3. Turn the lights off in the room, approach your bird and encourage it to step up onto the glove.

4. Mist your bird with a spray bottle and soak down its feath-

ers. If it tries to fly it won't be able to go far and can be more easily recovered.

If your bird is not reachable and/or the above techniques did not work, then you need to get a little more aggressive and actively pursue capturing your raptor. Here are a few strategies you can use:

1. If your bird is flighted, encourage it to fly back and forth to tire it out. If it gets exhausted, it may come lower into a position in which it can be retrieved.

2. If the bird can be reached using a ladder but won't stay still, turn the lights off to make the room dark and then proceed to get the bird on the glove or grab it.

3. Darken the room, turn a few flashlights on and encourage the bird to fly by making noise, throwing your glove in the air, etc. As your bird takes off, turn off the flashlights. Insecure in complete darkness, raptors tend to come lower. Flicking the light on and off will help your bird find a safe lower place to perch, hopefully within reach.

4. Turn the light off in the room, but leave an adjacent hallway door open with the light on. Flush your bird and it should move toward the light and into the hallway. Only do this if it will put your bird into a position to be more easily retrieved.

5. If all else fails, set a trap as described below. Usually, it does not come to this.

10.2 LOCATING A LOST BIRD

If your bird escapes outside, the best thing to do is stay calm. If you witness its departure, watch to see where it goes and perches, and have one person keep an eye on it while you either approach it or get help. If you lose sight of your bird, take a moment to collect yourself and think about your bird's abilities (How well can it fly?), the natural history of its species (Does it roost in evergreen trees? Is it a ground dweller?), and what resources are available to help you recover it. If your bird flies out of sight, note the direction in which it was headed and quickly round up a team of people to help.

Send people out in pairs, with two-way radios or cell phones if possible, so that if they spot the bird, you can be quickly notified and/or one person can stay with it while the other one gets help. Make sure you have quick access to handling gloves, raptor food (if your bird is fed on the glove) and a net so if you locate the bird you have some tools to try to capture it. Start at your bird's area of departure (or the last place you saw it) and cover the area in circles, moving farther from the original location as you go.

As mentioned earlier, unless your bird is a larger falcon or has perfect flight ability and catches a thermal, it probably didn't fly far. If it isn't capable of normal flight, keep in mind that it might perch somewhere low. If it's completely flighted, it will probably perch high. Make sure to check any evergreen trees in the area; some species that normally live in evergreens (long-eared owls, Northern saw-whet owls) might find and perch within them for cover. Also, large falcons generally won't fly into dense woodlots. Use your knowledge of the bird's natural history and individual behavior patterns to predict where and how it might perch.

Sometimes, the resident bird population will help you locate your avian educator. Listen for songbird and crow alarm calls and watch for crow mobbing behavior. These birds can often locate even the most elusive raptor.

10.2a Resources

If your initial search is fruitless, you might need to enlist the help of other people. Local falconers, bird-watchers, and residents, might all be willing to help. If the sun goes down and you still haven't found your bird, create a plan of action for the next day, starting at the crack of dawn from where the bird was last seen. Diurnal birds tend to take one last flight before dark to find a secure roosting spot and make their first flight of the day early when daylight breaks. Therefore, you need to be in position to search before first light.

At this point, if you do not know where your bird is, it is a good idea to expand your resources. The more people who know about your "jailbird" the better. Retrieving your bird safely should be your number one objective. It may be embarrassing to admit to your dilemma, but most people understand that things happen and will want to help. Resources to contact include:

- Your local wildlife office
- The humane society
- Your local police department
- Falconers
- Rehabilitators
- Zoos
- Your Audubon chapter
- Your state ornithological union list serve
- Local bird watching clubs

- Businesses in the area
- Animal control
- Local TV and radio stations

You can also post flyers in the general area where the bird disappeared. If your bird is fully flighted and you are located near the borders of other states, contact the wildlife offices of these states as well as rehabilitators and bird watching clubs. Give people clear instructions to contact you if the bird is sighted. Make sure they do this before approaching or trying to rescue the bird. They should only act (or not act) on your direction.

Don't give up hope. Even fully flighted birds can be found and recovered. If your bird left with jesses, swivel, and/or leash, it has the high probability of getting tangled in branches or other obstacles and will die if not recovered. This bird is your responsibility, and you need to do everything you can to bring it to safety. Therefore, don't quit looking for it after the first day. It may take many days and potentially weeks to recover your bird. If you don't search for it, you won't find it.

If you don't find your bird within a few days, you should notify your USFWS region. Federal regulations dictate that your region must be notified of a change in a bird's status within four days.

10.3 RECOVERING A RAPTOR

Once you've spotted your wayward bird, the next challenge is to capture it. A raptor that hasn't been trained to fly to your gloved hand and/or feed off the glove isn't going to come to you freely for food or companionship. Wild birds that are manned for educational use often quickly revert to a more wild state when in a natural habitat free of confinement. Therefore, don't expect your bird to be easily approached and to step onto the glove willingly. When free, a wild raptor's mindset often changes and its connection to you is broken, at least temporarily. Human imprinted birds are sometimes an exception. Be prepared to turn to other alternatives for capturing your raptor.

10.3a Option 1

The first option is to provide your bird opportunities to rescue itself. There are several things you can do to encourage your bird to place itself in positions to be retrieved.

Approach and retrieve
If the bird is in reach, you should try to approach and retrieve it.

Always move toward your bird smoothly and slowly. Crouching down as you move forward might seem less intimidating to your bird and therefore keep it calmer. Keep an eye on how it behaves as you approach. If it starts to look anxious, opens its wings in anticipation of flight, hops from branch to branch, or looks around intently for a place to go, stop moving and wait until it calms down. Try not to look directly at the bird if it is fidgety. Instead, move forward during times when the bird turns its head in a different direction. Move at a slow, steady pace, with no quick, jerky movements. Talk to your bird as you normally might and follow the same routine for offering food and/or the glove.

Offer branch to perch on
If your bird is hesitant to step up onto your gloved hand and keeps moving away from you, try offering a solid tree branch or dowel. Sometimes you can retrieve a raptor by encouraging it to step up onto a branch and then lifting the branch above your head as you move to an enclosed area, or a position that it can be grabbed if necessary.

Provide crate to crate-trained raptor
Another opportunity that you can provide for your bird will work if your bird is crate trained. Place the crate near the bird and leave a small amount of food on top and a larger portion inside. Back off and observe. If your bird is used to feeding in a crate, it may come down and enter for its meal.

10.3b Option 2

Actively pursue capturing your bird. If your bird does not take the opportunities presented to it to come home, you will need to turn to more active measures to recover it. Capturing your bird with a net or noose pole, retrieving your bird in the dark, trapping it with specific raptor traps, and soaking it with water are all methods that have been successfully used.

Capture a bird with a net or pole
If you try to approach your bird several times and are unable to get close enough to encourage it to step on the glove or to grab it, try capturing it with a fishing net or noose pole. A large fishing net with a long handle (expandable handles are great) can help you reach a bird that won't let you get close. A noose pole can also be used in a similar fashion. This is a long pole that has a retractable rope at the end to snare around wild animals. If you use it, slowly move the pole towards the bird and if possible place it near the bird's feet. If your bird remains calm, gently touch a foot to see if the bird will move and

place a foot in the middle of the noose. If you are lucky and it does, pull the end of the rope at the bottom of the handle to trap a leg.

You can also tie a large piece of food to a block on the ground. If your bird comes down to feed, its attention at times will be focused on the food and you can use this to move a net or pole closer. Patience is required. At this point, you just want to safely regain control of your bird; it will forgive you later for the method you use. Trapping your bird with a net or pole should only be attempted if the bird is in an easy position to retrieve. Do not attempt to use these tools if your bird is unreachable in a tree unless a professional climber is present and can scale the tree quickly once the bird is captured.

Retrieve a bird in the dark

Despite your best attempts, your bird might not stay still long enough to allow you to capture it using the above techniques. If that's the case, don't worry, just move on to another method. Many "escapees" can be caught at night in the dark (yes, even owls).

Once again, you must always know how the bird is acting as you proceed. If the bird is in a reachable position, it will usually allow you to approach to a reasonable distance (as compared to wild birds, which quickly flush). Talk to it as you approach. Your voice will be familiar and may prevent your bird from getting spooked and moving in the dark.

For this method, the most important piece of equipment is a bright flashlight. Once you reach the proximity at which your bird seems anxious, direct the flashlight at the bird's face and repeatedly turn it on and off at one-second intervals. This sudden light/dark pattern will confuse the bird as its eyes try to quickly adjust. Often this gives you enough time to approach the bird and grab it.

Trap a bird with conventional raptor traps

Bird banders, falconers, biologists, and wildlife officials have designed a variety of traps for safely capturing raptors of all sizes. Mist nets, bow nets, and bal-chatri traps are common types.

Mist net: (figure 10.1a) This long, fine net is supported by poles that you can either set into the ground or hold. (Birds are often spooky with people around, so setting the poles in the ground works best.) Once the net is set, you can encourage the bird to fly into it. The net is so fine that most birds don't even see it until they're trapped. You might need several nets to cover a large area.

Collapsible net: A small, thin flexible net can be set up vertically from the ground around a prey carcass or live prey item that is

Fig. 10.1
Devices commonly used to trap raptors.
(Gail Buhl)

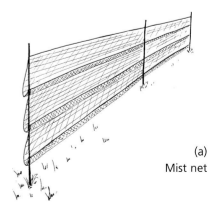

(a)
Mist net

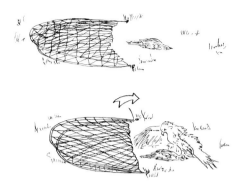

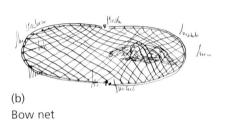

(b)
Bow net

(c)
Bal-chatri

securely fastened to a pole or block (to prevent your bird from carrying it away). The net's hole diameter should small enough so the bird can't easily pass through it (its head can but not its body), but large enough to distract the image so the bird does not identify it with a solid or suspicious object to avoid. When the bird approaches to investigate the food, the collapsible net will either get caught in a wing as the bird flaps, or completely cover a smaller bird, trapping it.

Like all the other traps, a net must be monitored constantly.

Bow net: (figure 10.1b) This spring-loaded net is set open with live prey attached on the inner side. When a bird gets hungry (or bored), it might respond to the movement of the prey and fly down to investigate or have a free meal. When it does, you can release the spring so the net quickly covers it. This trap must be attended constantly, but you should keep everyone hidden to avoid spooking the bird.

Bal-chatri: (figure 10.1c) This small wire cage (a square or half circle is the most common) is equipped with numerous fishing-line nooses over the sides and top. A few live mice (or other prey species) are placed inside, and their activity attracts the hungry raptor. When the bird tries to grab the prey, one or more of its toes gets trapped in the nooses.

It's important to use a bal-chatri trap heavy enough that the raptor can't carry it off. The size of the trap, therefore, will depend on the size of the bird. Once the bird is trapped, it must be retrieved immediately. Like the bow net, this type of trap must always be attended (you should be hidden but constantly watching it).

If you decide to try to trap your raptor, please contact a local bird bander (for mist nets), falconer, or wildlife biologist (for other types of traps) to help locate a trap and assist you with all the important details. Trapping a raptor in this fashion requires skill.

Soak a bird water

Okay. So you've tried everything feasible, your bird is moving a lot, and you don't have quick access to traps. What next?

If a raptor gets extremely wet, its temperament changes and it generally acts calmer. In addition, if it gets really drenched with water that has just a tinge of soap, it can't get much lift and flies poorly. (A very small amount of detergent changes the consistency of water and helps it to stay on the feathers; however, only a tiny amount is needed to sufficiently alter the water — the water should not appear soapy. Soapy water is irritating to a bird's eyes).

You can use these facts to your advantage. If possible, soak your bird, using a heavy misting stream of water. Don't use a coarse spray. A mist creates a fog that limits your bird's visual field and your bird may be more resistant to taking off. To spray your bird you can use a metal canister (like those used for spraying trees), a super-soaker squirt gun, or you can even enlist the assistance of your local fire department. Extreme situations sometimes call for extreme measures.

Climb a tree

You may notice that climbing a tree to retrieve your raptor has not been previously discussed. Although it is often people's first inclination to climb a tree, there are many factors to consider. First of all, during the daytime, most flighted birds won't stay in a tree while you climb to get it. Then, as it moves, you are totally out of position to act further. Enlisting a professional tree climber can help, although your bird is not familiar with this person and the probability that it will let this person retrieve it is slim. At night, it is possible to capture a diurnal bird by climbing its roosting tree, although many times the bird will be spooked with close movement at night and flush.

Another important consideration is your safety. If you are not an experienced climber, you could get seriously hurt. The only time when tree climbing is absolutely necessary is if your bird gets tangled in a tree by its equipment. However, if you are not comfortable climbing and the particular tree is not safe or the site is extremely high, recruit someone experienced to help you. Tree trimmers, some falconers, and even the fire department may be able to assist.

Do not cut the branch or tree down. This will undoubtedly result in injury to your bird as it plummets to the ground.

10.4 SUMMARY

If your raptor accidentally escapes from your control, stay calm and get organized. Try to keep an eye on where it goes and enlist help. If you lose sight of it, organize a search party, notify local wildlife organizations and residents, and post flyers. Once you find your bird, you can use a variety of techniques to retrieve it, including slow approaches, retrieval at night, setting traps, or soaking it with water. Don't give up hope. Recoveries may take several days or, depending on the situation, weeks.

10.5 SUGGESTED READINGS

Kimsey, B., and J. Hodge. 1992. *Falconry Equipment.* Houston: Kimsey/Hodge Publications.

Appendix A:
Federal and State Wildlife Permit Offices

AA.1 FEDERAL WILDLIFE PERMIT OFFICES

AA.1a Region 1

California
Hawaii
Idaho
Nevada
Oregon
Washington
American Samoa
Guam
Marshall Islands
Northern Mariana Islands
Trust Territory of the Pacific Islands

USFWS
Tami Tate-Hall
Migratory Bird Permit Office
911 N.E. 11th Ave.
Portland, OR 97232-4181
503-231-2019
Fax: 503-231-2019
tami_tatehall@fws.gov

AA.1b Region 2

Arizona
New Mexico
Oklahoma
Texas

Kamile McKeever
Migratory Bird Permit Office
P.O. Box 709
Albuquerque, NM 87103-0709
505-248-7885
Fax: 505-248-7885
kamile_mckeever@fws.gov

AA.1c Region 3

Illinois
Indiana
Iowa
Michigan
Minnesota
Missouri
Ohio
Wisconsin

Migratory Bird Office
BWH Federal Building
Fort Snelling, MN 55111
612-713-5393

AA.1d Region 4

Alabama
Arkansas
Florida
Georgia
Kentucky
Louisiana
Mississippi
North Carolina
South Carolina
Tennessee
Puerto Rico
Virgin Islands

Carmen Simonton
Migratory Bird Permit Office
P.O. Box 49208
Atlanta, GA 30359
404-679-4180
Fax: 404-679-4180
Carmen_simonton@fws.gov

AA.1e Region 5

Connecticut
Delaware
District of Columbia
Maine
Maryland
Massachusetts
New Hampshire
New Jersey
New York
Pennsylvania
Rhode Island
Vermont
Virginia
West Virginia

David Dobias
Migratory Bird Permit Office
P.O. Box 779
Hadley, MA 01035-0779
413-253-8641
Fax: 413-253-8424
david_dobias@fws.gov

AA.1f Region 6

Colorado
Kansas
Montana
Nebraska
North Dakota
South Dakota
Utah
Wyoming

Janell Suazo
Migratory Bird Permit Office
P.O. Box 25486
Denver, CO 80225-0486
303-236-5416
Fax: 303-236-8017
janell_suazo@fws.gov

AA.1g Region 7

Alaska

Meg Laws
Migratory Bird Permit Office
1011 E Tudor Rd
Anchorage, AK 99503
907-786-3693
Fax: 907-786-3641
meg_laws@fws.gov

**AA.2 STATE WILDLIFE PERMIT
OFFICES**

Alabama
Division of Wildlife/Freshwater
Fisheries
P.O. Box 301456
Montgomery, AL 36130-1456
334-242-3467

Alaska
Dept. of Fish and Game
P.O. Box 25526
Juneau, AK 99802-5526
907-465-6197

Arizona
Game and Fish Dept.
2221 W. Greenway Rd.
Phoenix, AZ 85023-4312
623-587-0139

Arkansas
Game and Fish Commission
31 Hallowel Lane
Humphrey, AR 72073
870-873-4302

California
Dept. of Fish and Game
3211 S. St.
Sacramento, CA 95816
916-227-1305

Colorado
Division of Wildlife
Special Licensing
P.O. Box 49128
Colorado Springs, CO 80919
719-268-0143

Connecticut
Dept. of Environmental Protection
Wildlife Division
79 Elm St.
Hartford, CT 06106-5127
860-424-3011

Delaware
Division of Fish and Wildlife
4876 Hay Point Landing Rd.
Smyrna, DE 19997
302-739-5297

Florida
Game and Wildlife Conservation
620 S. Meridian St.
Tallahassee, FL 32399-1600
850-488-6253

Georgia
Dept. of Natural Resources
2109 U.S. Hwy 278 SE
Social Circle, GA 30025
770-761-3044

Hawaii
Dept. of Land and Natural
Resources
Division of Forestry and Wildlife
1151 Punchbowl St.
Honolulu, HI 96813
808-587-0166

Idaho
Dept. of Fish and Game
600 S. Walnut
Boise, ID 83707-0025
208-334-2920

Illinois
Dept. of Natural Resources
One Natural Resource Way
Springfield, IL 62702-1271
217-782-7305

Indiana
Dept. of Natural Resources
402 W. Washington St. #W-273
Indianapolis, IN 46204-2212
317-232-4080

Iowa
Dept. of Natural Resources
502 E. Ninth St.

Wallace State Office Building
Des Moines, IA 50319-0034
515-281-8524

Kansas
Dept. of Wildlife and Parks
Fisheries and Wildlife Division
512 SE 25th Ave.
Pratt, KS 67124-8174
620-672-5911

Kentucky
Dept. of Fish and Wildlife
Resources
#1 Game Farm Rd.
Frankfort, KY 40601
502-564-3400

Louisiana
Dept. of Wildlife and Fisheries
P.O. Box 98000
Baton Rouge, LA 70898-9000
225-765-2976

Maine
Dept. of Inland Fisheries and
Wildlife
284 State St., Station #41
Augusta, ME 04333-0041
207-287-5240

Maryland
Dept. of Natural Resources
580 Taylor Ave.
Tawes Sate Office Building
Annapolis, MD 21401
410-260-8540

Massachusetts
Division of Fisheries and Wildlife
251 Causeway St.
Boston, MA 02114-2104
617-626-1590

Michigan
Dept. of Natural Resources
Wildlife Division
Box 30444
Lansing, MI 48909-7944
517-373-9329

Minnesota
Dept. of Natural Resources
500 Lafayette Rd.

St. Paul, MN 55115-4025
651-297-8040

Mississippi
Dept. of Wildlife, Fisheries, and
Parks
2148 Riverside Dr.
Jackson, MS 39202-1353
601-354-7303

Missouri
Dept. of Conservation
P.O. Box 180
Jefferson City, MO 65102-0180
573-751-4115, ext. 3262

Montana
Dept. of Fish, Wildlife, and Parks
1420 East Sixth Ave.
Helena, MT 59620-0701
406-444-2452

Nebraska
Game and Parks Commission
105 W. 2nd, Suite #201
Valentine, NE 69201
402-376-3116

Nevada
Dept. of Wildlife
Law Enforcement
1100 Valley Rd.
Reno, NV 89512
775-688-1500

New Hampshire
Fish and Game Dept.
11 Hazen Dr.
Concord, NH 03301
603-271-3127

New Jersey
Division of Fish, Game, and
Wildlife
P.O. Box 400
Trenton, NJ 08625-0400
609-292-2966

New Mexico
Dept. of Game and Fish
P.O. Box 25112
Santa Fe, NM 87504
505-476-8064

New York
Dept. of Environmental
Conservation
625 Broadway
Albany, NY 12233-4752
518-402-8985

North Carolina
Wildlife Management
Mail Service Center 1724
Raleigh, NC 27699
919-661-4872

North Dakota
State Game and Fish Dept.
100 N. Bismark Expressway
Bismark, ND 58501
701-328-6305

Ohio
Dept. of Natural Resources
Division of Wildlife
1840 Belcher Dr.
Columbus, OH 43224
614-265-6300

Oklahoma
Dept. of Wildlife Conservation
P.O. Box 53465
Oklahoma City, OK 73152
405-521-3719

Oregon
Dept. of Fish and Wildlife
3406 Cherry Ave. NE
Salem, OR 97303
503-947-6301

Pennsylvania
Game Commission
2001 Elmerton Ave.
Harrisburg, PA 17110-9797
717-783-8164

Rhode Island
Dept. of Environmental
Management
Division of Fish and Wildlife
Box 218
West Kingston, RI 02892
401-789-0281

South Carolina
Wildlife Permit Coordinator

Sandhills Research and Education
Center
P.O. Box 23205
Columbia, SC 29224-3205
803-419-9645

South Dakota
Game, Fish, and Parks Dept.
Division of Wildlife
523 E. Capitol Ave.
Pierre, SD 57501-3182
605-773-4229

Tennessee
Captive Wildlife Coordinator
TWRA/Law Enforcement Division
P.O. Box 40747
Ellington Ag. Center
Nashville, TN 37204
615-781-6647

Texas
Parks and Wildlife Dept.
4200 Smith School Rd.
Austin, TX 78744-3291
512-389-4481

Utah
State Dept. of Natural Resources
Division of Wildlife Resources
1594 W. North Temple
Suite 2110
Salt Lake City, UT 84114-6301
801-538-4701

Vermont
Agency of Natural Resources
Fish and Wildlife Dept.
103 S. Main, 10 South
Waterbury, VT 05671-0501
802-241-3727

Virginia
Dept. of Game and Inland Fisheries
Division of Permits
P.O. Box 11104
Richmond, VA 23230-1104
804-367-9588

Virgin Islands
Division of Fish and Wildlife
6291 Estate Nazareth 101
St. Thomas, VI 00802-1104
340-775-6762

Washington
Dept. of Fish and Wildlife
600 Capital Way N.
Olympia, WA 98501-1091
360-902-2513

West Virginia
Division of Natural Resources
1900 Kanawha Blvd.
Bldg. 3, Rm. 816
Charleston, WV 25305
304-558-2771

Wisconsin
Dept. of Natural Resources
Bureau of Wildlife Management
Box 7921
101 S. Webster St.
Madison, WI 53707-7921
608-264-6046

Wyoming
Game and Fish Dept.
Wildlife Division
5400 Bishop Bldg.
Cheyenne, WY 82006
307-777-457

Appendix B:
Food and Supplement Suppliers

Listed below are a few suppliers of raptor food. This list is not by any means exhaustive, and before choosing a supplier you should look into the quality of its products and handling methods. If there is not a supplier listed near you, check the web for additional sources.

AB.1 FOOD SUPPLIERS

AB.1a Rodents

Don's Rodents
P.O. Box 453
Wyoming, MN 55092
651-462-8973

Gourmet Rodent
6115 SW 137th Ave.
Archer, FL 32618
352-495-9024
Fax: 352-495-9781
www.gourmetrodent.com

Midwest Reptile and
Frozen Rodents
9665 N. Punkinvine Ct.
Fairland, IN 46126
317-861-5550
Fax: 317-861-5553
www.midwestreptile.com/
    rodents/html

Mike Dupuy Hawk Food
14405 New Hampshire Ave
Silver Spring, MD 20904
301-989-2222
Fax: 301-879-7880
www.mikedupuyhawkfood.com

Perfect Pets, Inc.
23180 Sherwood
Belleville, MI 48111
734-461-1362
1-800-366-8794
Fax: 734-461-2858
www.perfectpet.net

Rodents Plus
358 Wilkins Rd.
Mount Olive, NC 28365
919-658-9535
www.rodentsplus.com

Rodent Pro
P.O. Box 4424
Evansville, IN 47724
812-867-7598
Fax: 812-867-6058
www.rodentpro.com

The Big Cheese Rodent Factory
2527 W. Dickson St.
Fort Worth, TX 76110
1-800-887-0921
www.bigcheeserodents.com

AB.1b Coturnix Quail

Boyd's Bird Co.
3805 Airport Rd
Pullman, WA 99163
509-332-3109

Browning Quail
510 N. Edith
Shoshone, ID 83352
208-886-2898
www.calhawkingclub.org/vendors/
    BrowningQuail/Default.html

Cavendish Game Birds, Inc.
396 Woodbury Rd.
Springfield, VT 05156
802-885-1183
Fax: 802-885-5393
www.vermontquail.com

Invergo Quail Farm
7151 W. Blackjack Rd.
Hanover, IL 61041
815-777-2515

Johnson Quail
P.O. Box 1053
Owatonna, MN 55060
507-456-6192

Northwest Gamebirds
228812 E. Game Farm Road
Kennewick, WA 99337
509-586-0150
Fax: 509-585-9766
www.northwest-gamebirds.com

Morris Quail Farm Inc.
18370 SW 232 St.
Miami, FL 33170
305-247-1070
Fax: 305-247-0982
www.morrisinc.com

Rodent Pro
P.O. Box 4424
Evansville, IN 47724
812-867-7598
Fax: 812-867-6058
www.rodentpro.com

Rogue Valley Quail and Mice
13794 Perry Rd.
Central Point, OR 97502
541-826-1499
www.quailandmice.com

AB.1c Day-old Chicks

Local hatcheries

Mike Dupuy Hawk Food
14405 New Hampshire Ave
Silver Spring, MD 20904
301-989-2222
Fax: 301-879-7880
www.mikedupuyhawkfood.com

Morris Quail Farm Inc.
18370 SW 232 St.
Miami, FL 33170
305-247-1070
Fax: 305-247-0982

Rodent Pro
P.O. Box 4424
Evansville, IN 47724
812-867-7598
Fax: 812-867-6058
www.rodentpro.com

Johnson Quail
P.O. Box 1053
Owatonna, MN 55060
507-456-6192

AB.1d Fish

Local natural resources department,
fisheries, or commercial fisherman
(make sure you check into permit
requirements for fish acquisition
and storage, etc.)

AB.1e Commercial Diets

Nebraska Brand Bird of Prey Diet
Central Nebraska Packing, Inc.
P.O. Box 550
North Platte, NE 69103-0550
308-532-1250
1-800-445-2881
www.nebraskabrand.com

Toronto Feline Diet
Milliken Meat Wholesale
3447 Kennedy Road
Scarborough, ON
Canada M1V 3S1
416-299-9600

AB.2 VITAMIN SUPPLEMENTS

AB.2a Cod Liver Oil (Vitamins A and D)

UPCO
3705 Pear St.
St. Joseph, MO 64501
1-800-254-8726

Local pharmacies and drug stores

AB.2b Nutrical

UPCO
3705 Pear St.
St. Joseph, MO 64501
1-800-254-8726

Local veterinarians

AB.2c Thiamine (Vitamin B Complex)

Local pharmacies and drug stores
Local veterinarians

AB.2d Vitahawk (Multivitamin)

D.B. Scientific
2063 Main St., Suite 406
Oakley, CA 94561
vitahawk@gmail.com
www.vitahawk.com

Mike's Falconry Supplies
4700 SE Chase Rd.
Gresham, Oregon, 97080
1-888-663-5601
mikes@mikesfalconry.com
www.mikesfalconry.com

Northwoods, Limited
P.O. Box 874
Ranier, WA 98576
360-446-3212
www.northwoodsfalconry.com

AB.2e Vionate (Multivitamin)

UPCO
3705 Pear Street
St. Joseph, MO 64501
1-800-254-8726

Local veterinarians

Appendix C: Raptor Housing

Audubon of Florida
Center for Birds of Prey
1101 Audubon Way
Maitland, FL 32751
407-644-0190
www.audubonofflorida.org

Figure AC-1a,b: Inside views of kite display
Figure AC-2: Inside view of medium bird enclosure
Figure AC-3: Inside view of bald eagle enclosure
Figure AC-4: Inside view of carcara/vulture display

Back to the Wild
Wildlife Rehabilitation and Nature Education Center
P.O. Box 423
4504 Bardshar Rd.
Castalia, Ohio 44824
419-684-9539
www.backtothewild.com

Figure AC-5: Outside view of several adjacent
    enclosures
Figure AC-6: Frontal view of bald eagle enclosure
Figure AC-7: Eastern screech owl enclosure
Figure AC-8: Inside view of large hawk or owl
    enclosure

Last Chance Forever
311 E. North Loop Rd.
P.O. Box 460993
San Antonio, TX 78216
210-499-4080

Figure AC-9: Frontal view of multiple mews
    surrounding a flight enclosure
Figure AC-10: Outside view of crested caracara
    enclosure
Figure AC-11: Inside view of hawk enclosure
Figure AC-12: Angled view of red-tailed hawk enclosure

Native Bird Connections
Diana Granados
PMB 156-6680 Alhambra Ave.
Martinez, CA 94553-6105
925-947-7044
www.nativebirds.org

Figure AC-13: Outside view of four adjacent enclosures
Figure AC-14: Outside view of single enclosure
Figure AC-15: Inside view of red-tailed hawk enclosure
Figure AC-16: Inside view of prairie falcon enclosure
Figure AC-17: Vestibule hallway for adjacent enclosures
Figure AC-18: Inside view of turkey vulture enclosure

Raptors of the Rockies
P.O. Box 250
Florence, MT 59833
406-829-6436
www.raptorsoftherockies.org

Figure AC-19: Outside view of hawk enclosure
Figure AC-20: Inside view of great horned enclosure
Figure AC-21: Outside view of medium-sized raptor
    enclosure

The Raptor Center
College of Veterinary Medicine
University of Minnesota
1920 Fitch Ave.
St. Paul, MN 55108
612-624-4745
www.theraptorcenter.org

Figure AC-22: Bald eagle display
Figure AC-23: Red-tailed hawk display
Figure AC-24: Bald eagle enclosure
Figure AC-25: American kestrel enclosure
Figure AC-26: Peregrine falcon enclosure

The Raptor Trust
1390 White Bridge Road
Millington, NJ 07946
908-647-2353
www.theraptortrust.org

Figure AC-27: Diagram of a standard raptor exhibit
    enclosure
Figure AC-28: Outside view of hawk enclosure
Figure AC-29: Inside view of hawk enclosure
Figure AC-30: Inside view of falcon enclosure

Thomas Irving Dodge Nature Center
365 Marie Ave. W.
West St. Paul, MN 55118
651-455-4531
www.dodgenaturecenter.org

Figure AC-31: Front view of four adjacent enclosures
Figure AC-32: Front view of red-tailed hawk enclosure
Figure AC-33: Back window of red-tailed hawk
    enclosure
Figure AC-34: Inside of bald eagle enclosure

Vermont Institute of Natural Science
Vermont Raptor Center
27023 Church Hill Rd.
Woodstock, VT 05091-9642
802-457-2779
www.vinsweb.org

Figure AC-35: Angled view of red-tailed hawk
    enclosure
Figure AC-36: Front view of great horned owl
    enclosure
Figure AC-37: Inside of long-eared owl enclosure
Figure AC-38: Inside of Northern saw-whet owl
    enclosure
Figure AC-39: Inside of peregrine falcon enclosure
Figure AC-40: Inside of golden eagle enclosure

Figure AC-1a

Figure AC-1b

Figure AC-2

Figure AC-3

Figure AC-4

Figure AC-5

Figure AC-6

Figure AC-7

Figure AC-8

Figure AC-9

Figure AC-11

Figure AC-12

Figure AC-10

Figure AC-13

Figure AC-14

Figure AC-15

Figure AC-16

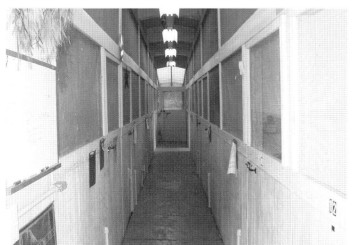

Figure AC-17

Figure AC-18

Figure AC-19

Figure AC-20

Figure AC-21

Figure AC-22

Figure AC-23

Figure AC-24

Figure AC-25

Figure AC-26

Figure AC-27

4'

8'

24'

16'

Figure AC-28

Figure AC-29

Figure AC-30

Figure AC-31

Figure AC-32

Figure AC-33

Figure AC-34

Figure AC-35

Figure AC-36

Figure AC-37

Figure AC-38

Figure AC-40

Figure AC-39

Appendix D:
Handling and Training Supplies

AD.1 EQUIPMENT AND SUPPLY CATALOGS

Falconry equipment vendors
www.falconry.com

Mike's Falconry Supplies
4700 SE Chase Rd.
Gresham, Oregon, 97080
1-888-663-5601
mikes@mikesfalconry.com
www.mikesfalconry.com

Northwoods Limited
P.O. Box 874
Rainier, WA 98576
1-800-446-5080
www.northwoodsfalconry.com

Nuzzo Raptor Equipment
5695 W. Hill Rd.
Decatur, IL 62522
217-963-6909
nuzzoraptorequipment@juno.com

AD.2 SPECIALTY SUPPLIERS

AD.2a Books and Videos (catalogs available)

Buteo Books
3130 Laurel Rd.
Shipman, VA 22971
1-800-722-2460
www.buteobooks.com

Falconry equipment vendors
www.falconry.com

Hancock House Publishers
1431 Harrison Ave
Blaine, WA 98230-5005
604-538-1114
1-800-938-1114
www.hancockhouse.com

Mike's Falconry Supplies
4700 SE Chase Rd.
Gresham, Oregon, 97080
1-888-663-5601
mikes@mikesfalconry.com
www.mikesfalconry.com

Minnesota Department of Natural
Resources Nongame Wildlife
Program Section of Wildlife
500 Lafayette Rd.
St. Paul, MN 55155
651-296-6157
www.dnr.state.mn.us

Northwoods Limited
P.O. Box 874
Rainier, WA 98576
1-800-446-5080
www.northwoodsfalconry.com

Western Sporting Publications
P.O Box 939
Ranchester, WY 82839-0939
307-672-0445
www.westernsporting.com

Wild Ones
Animal Health Library
P.O. Box 947
Springville, CA 93265-0947
1-800-539-0210

AD.2b Gloves

Falconry equipment vendors
www.falconry.com

Local welding supply stores,
liquidators, hardware stores
(welders gloves and small
leather gloves)

Mike's Falconry Supplies
4700 SE Chase Rd.
Gresham, Oregon, 97080
1-888-663-5601
mikes@mikesfalconry.com
www.mikesfalconry.com

Northwoods Limited
P.O. Box 874
Rainier, WA 98576
1-800-446-5080
www.northwoodsfalconry.com

One-of-a-Kind
327 E. Lake Street
Horicon, WI 53022

Pineo Falconry Supplies
4210 S. Dorset Rd.
Spokane, WA 99224
509-838-3621
www.pineofalconry.com

Traditions Glove
Jim Spohn
30469 South Ditmore Rd.
Worley, Idaho 83876
208-686-1936

AD.2c Hoods

Falconry equipment vendors:
www.falconry.com

Mike's Falconry Supplies
4700 SE Chase Rd.
Gresham, Oregon, 97080
1-888-663-5601
mikes@mikesfalconry.com
www.mikesfalconry.com

Northwoods Limited
P.O. Box 874
Rainier, WA 98576
1-800-446-5080
www.northwoodsfalconry.com

Nuzzo Raptor Equipment
5695 W. Hill Rd.
Decatur, IL 62522
217-963-6909
nuzzoraptorequipment@juno.com

Rick Woods
205 Wapsie Access Blvd.
Independence, IA 50644
RichardWoods@woodsfalconfarm.com

Pineo Falconry Supplies
4210 S. Dorset Rd.
Spokane, WA 99224
509-838-3621
www.pineofalconry.com

AD.2d Leather and Leather Supplies

Falconry equipment vendors:
www.falconry.com

Mike's Falconry Supplies
4700 SE Chase Rd.
Gresham, Oregon, 97080
1-888-663-5601
mikes@mikesfalconry.com
www.mikesfalconry.com

Northwoods Limited
P.O. Box 874
Rainier, WA 98576
1-800-446-5080
www.northwoodsfalconry.com

Tandy Leather Company
Stores in AZ, CA, CO, CT, FL, GA,
ID, IL, IN, KY, MD, MI, MN, MO,
NE, NV, NM, NY, OH, OK, PA,
TN, TX, VT, WA, VG, WI, and
Canada
www.tandyleather.com

AD.2e Perches

Falconry equipment vendors
www.falconry.com

Mike's Falconry Supplies
4700 SE Chase Rd.
Gresham, Oregon, 97080
1-888-663-5601
mikes@mikesfalconry.com
www.mikesfalconry.com

Northwoods Limited
P.O. Box 874, Rainier, WA 98576
1-800-446-5080
www.northwoodsfalconry.com

AD.2f Perch Covering Materials

Mike's Falconry Supplies
4700 SE Chase Rd.
Gresham, Oregon, 97080
1-888-663-5601
mikes@mikesfalconry.com
www.mikesfalconry.com
(Monsanto Astroturf, neoprene,
stadium mat/Astroturf)

Northwoods Limited
P.O. Box 874
Rainier, WA 98576
1-800-446-5080
www.northwoodsfalconry.com
(Monsanto Astroturf and stadium
mat/Astroturf)

Solutia Inc.
2381 Centerline Industrial Drive
St. Louis, MO 63146-3323
1-800-325-4330
Fax: 314-997-8652
www.solutia.com
(Monsanto Astroturf)

Three-strand Manilla rope: home
improvement stores, hardware
stores

AD.2g Scales

Berne Scale
2200 Edgewood Ave. S.
Minneapolis, MN 55426
952-544-2422
1-888-544-7835
www.bernescale.com

Falconry equipment vendors:
www.falconry.com

Northwoods Limited
P.O. Box 874
Rainier, WA 98576
1-800-446-5080
www.northwoodsfalconry.com

Mike's Falconry Supplies
4700 SE Chase Rd.
Gresham, Oregon, 97080
1-888-663-5601
mikes@mikesfalconry.com
www.mikesfalconry.com

Setra Scales
159 Swanson Rd.
Boxborough, MA 01719-1304
1-800-257-3872
www.setra.com

Sterling Scale
20950 Boening Dr.
Southfield, MI 48075
1-800-331-9931
www.sterlingscale.com

AD2.h Transmitter Equipment

Falconry equipment vendors:
www.falconry.com

LL Electronics
c/o Louis Luksander
P.O. Box 420
Mahomet, IL 61853
1-800-553-5328

Holohil Systems Ltd.
112 John Cavanaugh Dr.
Carp, Ontario
Canada, KOA 1LO
613-839-0676
Fax: 613-839-0675
www.holohil.com

Marshall Telemetry
896 W. 100 N
North Salt Lake
UT 84054
1-800-729-7123
Fax: 801-936-0900
www.marshallradio.com

Merlin Systems, Inc.
P.O. Box 190257
Boise, ID 83719
208-362-2254
Fax: 208-362-2140
www.merlin-systems.com

Mike's Falconry Supplies
4700 SE Chase Rd.
Gresham, Oregon, 97080
1-888-663-5601
mikes@mikesfalconry.com
www.mikesfalconry.com

Northwoods Limited
P.O. Box 874
Rainier, WA 98576
1-800-446-5080
www.northwoodsfalconry.com

U.S. Geological Survey
Radio Telemetry Equipment
Suppliers
www.npwrc.usgs.gov

Wildlife Materials, Inc.
Route 1
Carbondale, IL 62901
618-549-6330

Appendix E:
Maintenance and Medical Supplies

AE.1 Bandaging Materials
(available from a veterinarian or drugstore)

- Masking tape, Durapore silk tape (3M), or other relatively non-sticky tape
- Telfa pads
- Vet wrap (3M) or nylon bandaging material
- Mircofoam Tape (3M)

AE.2 Beak and Talon Care

- Clear fingernail polish
- Dog and/or cat nail trimmer (available from pet store)
- Emery boards/fingernail file
- Eyedropper
- Gauze (available from drug store)
- Kwik Stop® antiseptic powder (available from pet store)
- Rotary tool (available from home improvement stores, tool suppliers)
- Rotary bits (available from home improvement stores, tool suppliers)
- Plastic syringe case: 1cc (1ml) and 3cc (3ml)

AE.3 Control of Ectoparasites
(Parasite sprays/powders)

- Scalex ® Mite and Lice Spray — active ingredients include pyrethrins and pyperonyl butoxide (available from pet store)
- Sectrol® (3M) — active ingredient 2% pyrethrins (available from pet store)
- Sevin® — active ingredient carbaryl (available from plant nurseries, garden shops)
- Frontline® spray — active ingredient 7-9% Fipronil (available from veterinarian)

AE.4 Emergency Kit

- Antibiotic ointment — triple antibiotic ointment, betadine ointment (available from drug store)
- Emergency phone numbers
- Fluids — requires syringe and gavage tube if caretaker knows tubing procedure (see below)
- Gauze or cotton
- Iodine solution (available from drug store)
- Kwik Stop® antiseptic powder (available from pet store)
- Plastic 1cc or 3cc syringe case
- Spray bottle
- Vet wrap (3M), nylon bandaging material, and/or masking tape (available from drug store)

AE.5 Fluids

- Mammalian lactated ringers ("Viaflex"), Abbott Laboratories (available from a veterinarian)
- Oral electrolyte — Pedialyte® (available from drug store) Resorb® (Pfizer; available from a veterinarian)
- Boiled Coca-Cola

AE.6 Foot Care

- A and D ointment (available from drug stores)
- Aloe Vera Gelly® (99%) (Lily of the Desert, Denton, Texas 76208, 1-800-439-5506; available at drug stores)
- Aquaphor® Healing Ointment (Eucerin product, available at drug stores)
- Calendula, Echinacea, Hypericum Cream (available from "Quintessence," 334 W. Lakeside St., Madison, WI 53715; 608-251-6915
- Cotton-tipped applicators
- Newskin® liquid bandage (available from drug stores)
- Protecta-pad® (Tomlyn Products, 101 Lincoln Ave., Buena, NJ, 08310, 1-800-866-5582; available from some pet stores)
- Tuff-foot® (Bonaseptic Co., Atlanta, Georgia, 1-800-Tuf-foo; available from some pet stores)
- Udder balm (available from drug store or pet store)
- Natural emu oil (available from a variety of sources listed on the web)

AE.7 Weight Determination
(see appendix D)

- Scale (with perch for manned birds, without perch for unmanned birds)
- Towel or hood to cover bird's eyes if necessary

AE.8 Wound Care
(for temporary pre-veterinary care, available from drug stores or pet stores)

- Antibiotic ointment (triple antibiotic ointment, betadine ointment)
- Iodine solution, sterile saline
- Telfa pads
- Vet wrap (3M)
- Bandaging tape
- Sterile lubricant (petroleum jelly)
- Tegaderm (3M) transparent dressing

GLOSSARY

anklet: A leather strap fastened around the lower leg of a raptor by a grommet, through which an aylmeri jess passes.

aspergillosis: A potentially fatal fungal infection of the respiratory system, found in a wide variety of animals.

aylmeri: A type of jess system that consists of two pieces: an anklet fastened around the lower leg, and a strap that passes through a grommet in the anklet.

bate: A tethered raptor's attempt to fly off a gloved hand or perch.

blood feather: A growing feather, characterized by blood in the base of the shaft.

bridge: A sound or action that immediately tells an animal it performed a desirable behavior; it fills the gap of time between the act and positive reinforcement.

bumblefoot: Any degenerative or inflammatory condition of a bird's foot.

candida: A yeast infection of the gastrointestinal tract, found in many animals.

capallaria: An endoparasite that invades the gastrointestinal tract of birds.

cast: (1) To grab a raptor and place it on its back. (2) To regurgitate the indigestible components of a meal in the form of a pellet.

centimeter (cm): A metric unit equaling 0.4 inches.

coccidia: A protozoan endoparasite that invades a bird's liver, digestive system, kidneys and other tissues.

coping: Trimming and reshaping a raptor's beak and talons.

crop: An out pocketing of the esophagus in eagles, hawks, osprey, and vultures (not owls) that acts as a temporary storage organ for food.

cue: A sound or act that tells an animal what behavior is expected.

deck feathers: The two central tail feathers, which are slightly elevated from the others and often show a slightly different pattern and coloration.

demonstration bird: A raptor trained for free-flight programs.

display bird: A raptor that is housed in a facility to be viewed by the public; often not a trained (manned) bird.

ectoparasite: A parasite that feeds on the feathers or skin.

endoparasite: A parasite that invades the internal systems of an animal.

enrichment: Visual or tactile forms of desirable stimulation added to a captive raptor's environment to enhance its quality of life.

fisting: Asking a raptor to step up onto a gloved hand.

free-lofted: Allowed to fly freely within an enclosure.

gram (g): A metric unit equaling 1/28 of an ounce.

hackles: Feathers on the head and neck of hawks and eagles that are often raised as a defensive or aggressive warning.

hemoparasites: Parasites that invade the blood.

hock joint: The ankle joint of a raptor, separating its tarsometatarsus bone and tibiotarsal bone.

hood: A close fitting "cap" placed over the head and eyes of an excitable hawk or eagle to keep it calm during exams or transport.

imping: The process of replacing broken or damaged feathers.

imprinting: The process of orientation that occurs during the first days to weeks of life (depending on the species); a youngster usually orients on the species by which it was raised. Different types include parental, sibling, sexual, fear response, and habitat.

jess: A leather or nylon strap attached to the lower leg of a raptor to restrain it.

leash: A leather or nylon strap attached to a bird's jesses to tether it to a perch or restrain it on a gloved hand.

lure: An object, often a soft, handmade reproduction of a bird, that a free-flighted raptor is trained to fly to.

man: To teach a raptor to feel secure while being positively handled (fisted, crated, walked).

mantle: The behavior of spreading wings and tail over food to protect it.

meter (m): A metric unit equaling 3.3 feet.

mew: A raptor housing area that provides protection from the weather.

mute: The urates and fecal material of a raptor; excreted through its vent as a white circle (urates) with a dark center (fecal matter).

operant conditioning: A type of learning in which a desired behavior is associated with positive reinforcement.

parasite: An organism that lives in or on another organism and has the potential to cause it some degree of harm.

pellet: A small, compact bundle of indigestible food material (bones, feathers, fur) regurgitated by a raptor after it digests a meal.

positive reinforcement: Something desirable, such as food, presented immediately following a behavior that increases the likelihood that the behavior will occur again.

program bird: A raptor that has been trained to a gloved hand and is used in educational programs.

serratospiculum: Roundworms that invade the air sacs, especially in prairie falcons.

swivel: A stainless steel piece of equipment consisting of two rings (one stationary and one that rotates) that attach a raptor's jesses to its leash.

talon sheath: Hard outer covering (usually black) of a raptor's talon.

tarsometatarsus: The leg bone extending from the hock (ankle joint) to the foot.

tethered: Secured by a leash to a perch or gloved hand.

trochomonas: A protozoan parasite infection that targets the mouth and digestive system; often seen in raptors fed pigeons or other bird species that carry the parasite.

waking: A training method used to man raptors that involves keeping a raptor on a gloved hand for 24 consecutive hours or more.

weathering area: An outdoor housing area that provides a raptor access to sun.

zoonotic disease: A disease that can be transmitted from one animal to another.

INDEX

Italicized page numbers denote illustrations/photos

HANCOCK HOUSE PUBLISHERS

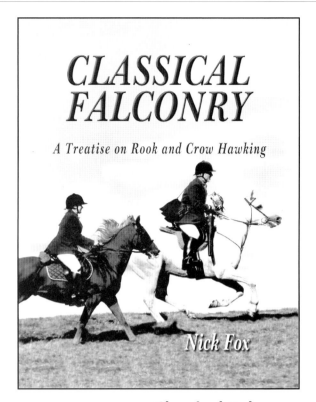

**Captive Raptor
Management & Rehabilitation**
Richard Naisbitt & Peter Holz
ISBN 0-88839-490-X
8½ x 11, HC, 176 pages

Classical Falconry
A Treatise on
Rook and Crow Hawking
Nick Fox
ISBN 0-88839-548-5
8½ x 11, HC, 248 pages

**Understanding the
Bird of Prey**
Nick Fox
ISBN 0-88839-317-2
8½ x 11, HC, 375 pages

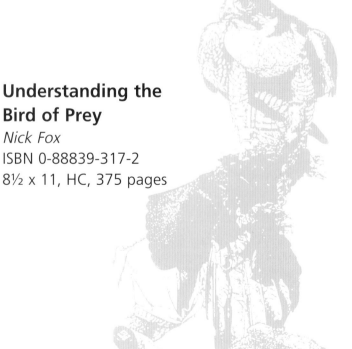

View all **HANCOCK HOUSE** raptor titles at **www.hancockhouse. com**

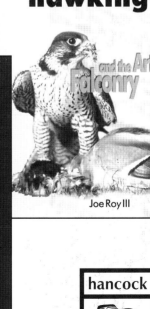

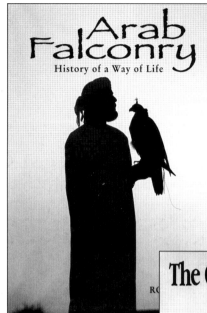

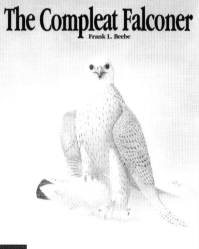
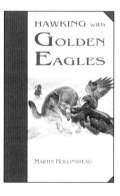
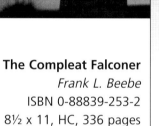